U.S. Marine Corps Helicopter borne Operations

Department of Defense

Pentagon
Publishing

To Our Readers

Changes: Readers of this publication are encouraged to submit suggestions and changes that will improve it. Recommendations may be sent directly to Commanding General, Marine Corps Combat Development Command, Doctrine Division (C 42), 3300 Russell Road, Suite 318A, Quantico, VA 22134-5021 or by fax to 703-784-2917 (DSN 278-2917) or by E-mail to **nancy.morgan@usmc.mil**. Recommendations should include the following information:

- Location of change
  - Publication number and title
  - Current page number
  - Paragraph number (if applicable)
  - Line number
  - Figure or table number (if applicable)
- Nature of change
  - Add, delete
  - Proposed new text, preferably double-spaced and typewritten
- Justification and/or source of change

Additional copies: A printed copy of this publication may be obtained from Marine Corps Logistics Base, Albany, GA 31704-5001, by following the instructions in MCBul 5600, *Marine Corps Doctrinal Publications Status.* An electronic copy may be obtained from the Doctrine Division, MCCDC, world wide web home page which is found at the following universal reference locator: **https://www.doctrine.usmc.mil**.

Unless otherwise stated, whenever the masculine gender is used, both men and women are included.

DEPARTMENT OF THE NAVY
Headquarters United States Marine Corps
Washington, D.C. 20380-1775

16 August 2004

FOREWORD

Marine Corps Warfighting Publication (MCWP) 3-11.4, *Helicopterborne Operations*, describes how infantry and aviation units plan and conduct helicopterborne operations during subsequent operations ashore. MCWP 3-11.4 emphasizes the coordination necessary between ground, air, combat support, and combat service support organizations concerning the planning sequence and tactical employment of ground and aviation elements. It describes the versatility of helicopterborne operations and explains the tactical fundamentals of helicopterborne operations for ground-based operations once ashore.

MCWP 3-11.4 is intended for commanders, staff officers, and support units responsible for the planning and execution of helicopterborne operations. However, it should be read by any Marine involved in the execution of helicopterborne operations.

This publication does not contain information relative to amphibious operations. Helicopterborne operations in amphibious operations are discussed in MCWP 3-31.5, *Ship-to-Shore Movement*, Naval Warfare Publication (NWP) 3-22.5, *Tactical Manual* series, and Joint Publication (JP) 3-02.1, *Joint Doctrine for Landing Force Operations*.

This publication supersedes Fleet Marine Force Manual (FMFM) 6-21, *Tactical Fundamentals of Helicopterborne Operations*, dated June 1991.

Reviewed and approved this date.

BY DIRECTION OF THE COMMANDANT OF THE MARINE CORPS

EDWARD HANLON, JR.
Lieutenant General, U.S. Marine Corps
Commanding General
Marine Corps Combat Development Command

Publication Control Number: 143 000147 00

TABLE OF CONTENTS

Chapter 4. Planning

Chapter 5. Combat Operations

Appendices

CHAPTER 1

OVERVIEW

Helicopterborne operations are tactical operations in which assault forces maneuver on the battlefield under the direction of an assigned commander in order to engage and destroy enemy forces or to seize key terrain. They are best employed in situations that provide the force a calculated advantage due to surprise, terrain, threat, or mobility. Helicopterborne operations allow the commander to maneuver rapidly to achieve tactical surprise and mass forces, regardless of obstacles and without dependency on ground lines of communication. These operations embody the combined-arms concept through coordination and planning between the air and ground commanders. Infantry and air units can be fully integrated with other members of the combined-arms team to form powerful and flexible helicopterborne task forces (HTFs). These forces can project combat power throughout the entire depth, width, and breadth of the modern battlefield with little regard for terrain barriers. The unique versatility and strength of an HTF is achieved by combining the capabilities of helicopters—speed, agility, and firepower—with those of the infantry and other combat arms to form tactically tailored HTFs that can be employed in low, medium, and high intensity environments.

Helicopterborne operations are not merely the movement of Marines, weapons, and material by helicopter units. They are deliberate, precisely planned, and aggressively executed combat operations that allow friendly forces to strike over terrain barriers in order to attack the enemy when and where he is most vulnerable.

Note

Helicopterborne operations are more than just air movement operations. Air movement operations require airlift assets outside of the assets used for helicopterborne operations. These operations are used to move troops and equipment; to emplace artillery and other combat support assets; and to transport ammunition, fuel, and supplies.

Helicopterborne operations span the spectrum of risk. They can be high-payoff operations that can, when properly planned and aggressively executed, drastically extend a commander's area of operations. This extension of a commander's area of operations enables the commander to execute operations in areas ranging beyond the capabilities of ground forces.

The speed and mobility of helicopters can provide freedom of rapid maneuver. Freedom of rapid maneuver can then fix the enemy and mass sufficient combat power to destroy him over distances that would otherwise be impossible to traverse as quickly. The helicopter's flexibility and versatility permit the ground commander to reduce time and distance limitations normally encountered in the ground movement of troops.

Sequence of Operations

A helicopterborne operation generally takes place in the following sequence:

- Planning.
- Briefing.
- Loading.
- Air assault.
- Landing.
- Tactical ground operations.
- Sustainment.
- Ground linkup/air reposition.

Tactical Considerations

In addition to utilizing the tactical fundamentals of ground combat, helicopterborne operations also apply the following tactical considerations:

- The HTF is assigned missions that take advantage of its superior mobility, and it is not employed in roles requiring deliberate operations over an extended period of time.
- The HTF fights as a combined-arms team.
- Operational planning must be centralized and precise.
- Execution must be aggressive and decentralized.
- Helicopterborne forces lack tactical mobility and heavy weapons; therefore, it is important that the force lands on or near the objective. The successful accomplishment of the mission can be threatened if the force lands further from the objective than planned, particularly if the enemy has superior ground mobility. Consequently, the landing of a helicopterborne force in any location other than the designated landing zone (LZ) is justified only when landing in the designated zone poses a threat to force survival.
- Helicopterborne forces may operate in conjunction with other ground forces or independently. Helicopterborne forces enable the commander to react quickly over the entire depth and width of his area of responsibility.
- Helicopterborne operations require a rapid buildup of combat power on the ground; e.g., one-third of the helicopterborne force's assault elements should be landed in the initial wave (for a battalion landing, a minimum of one company and for a company landing, a minimum of one platoon). The actual size of the assault wave is based on the threat and determined by the mission commander.
- Helicopters are an excellent means of tactical deception. When possible, helicopters can make demonstration landings in several different zones during one flight to deceive the enemy as to the true objective of an operation.
- Helicopterborne attacks are typically launched against undefended or lightly defended objectives. If attacking a well-defended objective, planners must select LZs that are nearby and support a safe landing of the force while adequately suppressing enemy air defenses.
- Helicopterborne forces are vulnerable to attack helicopters, fixed-wing aircraft, surface-to-air missiles, and antiaircraft artillery. To counter this vulnerability, fixed-wing and/or rotary-wing aircraft escort the helicopterborne forces, and use indirect fires to suppress or neutralize enemy air defenses. Suppression may also be conducted by electronic attack.
- During a landing, helicopterborne forces are especially vulnerable and may be disorganized for a short time.
- Air defense weapons are employed to counter this vulnerability in the pickup zones (PZs) and the LZs.
- Helicopterborne forces should be employed early to allow an early linkup with vehicular support to enhance ground mobility and sustainability. It is advisable to use the helicopterborne force as the stationary force during the linkup because of itslimited mobility while on the ground.
- The increased mobility of a helicopterborne force allows the commander to extend his area of influence, but the following risks must be considered:
  - Greater exposure to enemy ground fire and enemy aircraft.
  - Possible loss of surprise.
  - Possible involvement with enemy reserves.
  - Increased vulnerability to enemy counterattack pending linkup with ground forces.
  - Decreased lines of communications.
- To execute successful deep helicopterborne operations, the following must exist:
  - Detailed intelligence and objective area collection assets.
  - Ability to move securely to the objective area
  - Ability to execute the mission without ground lines of communications

  - Ability to provide aerial combat support and combat service support (CSS).
  - Ability to withdraw forces if required.
- Due to the lack of vehicles and other support during the initial stages of the helicopterborne operation, Marines must carry all necessary supplies and equipment; therefore, commanders must ensure that only mission-essential items are carried. The commander ensures that other supplies and equipment are delivered at the earliest opportunity.
- Coordination is required between ground and helicopter units involved in a helicopterborne operation; therefore, rehearsals are critical to mission success. At a minimum, communications and actions in the PZs and LZs are rehearsed.
- Command and control (C2) is tailored for the assigned mission. Ideally, the air mission commander (AMC) and helicopterborne unit commander (HUC) are collocated in a C2 platform. This collocation may not always be possible depending on the mission.
- Due to the range of helicopterborne operations, the use of airborne relay and C2 platforms must be considered and the communications plan must be simple, redundant, and fully integrate all elements of the force. See Marine Corps Warfighting Publication (MCWP) 3-40.3, *Communications and Information Systems*, appendix D, for a description of Marine air-ground task force (MAGTF) radio nets that can be used to provide the needed C2 links for the helicopterborne force.
- Helicopterborne operations conducted during adverse weather require more planning and preparation time.
- A unit's tactical integrity should be maintained throughout a helicopterborne operation.
- Fire support planning must provide for suppressive fires along flight routes and in the vicinity of LZs. The suppression of enemy air defenses (SEAD) systems must be a priority for fires.
- Helicopterborne forces are most effectively employed in environments where limited lines of communications are available to the enemy, terrain limits the use of heavy ground forces, and where the enemy lacks air superiority and effective air defense systems.
- When infantry units conduct helicopterborne operations, the HUC must determine the disposition of unit vehicles, attached vehicles, and support vehicles. Vehicles may be—
  - Flown in with the helicopterborne force.
  - Flown in subsequent to arrival of the helicopterborne force.
  - Driven in by ground mobile elements of the helicopterborne force subsequent to the initial assault.
  - Attached to another ground unit (e.g., a link-up force) for movement to the objective area.
  - Left in the assembly area until the helicopterborne force returns.
  - Staged in position to provide support for the helicopterborne force or adjacent units.

Also see appendix A for the small-unit leader's guide to tactical PZ/LZ operations.

Command and Control

Helicopterborne operations require close coordination between the commander of the ground unit to be lifted and the AMC. The AMC is an experienced naval aviator typically operating from an aircraft to direct airborne coordination and control of helicopterborne operations. When no AMC has been designated, the assault flight leader (AFL) performs this function within his capabilities. The following coordination measures enhance command and control of helicopterborne operations:

- Coordination begins at the earliest opportunity in the planning phase of the operation.
- When possible, both commanders are collocated, normally via a C2 helicopter, during the air movement and initial stages of the landing.

- The AMC's primary responsibility is to coordinate the air movement of personnel and equipment into designated LZs. The AMC supports the ground commander's concept of operations.
- While the air movement phase is primarily the responsibility of the AMC, the ground commander must be prepared to recommend primary and alternate approach and retirement lanes. The ground commander must confirm the proper LZ.
- It may become necessary to shift from primary to alternate LZs or to alter the course of helicopter flights; therefore, the authority to change to an alternate LZ must be established as soon as LZs are selected. The shifting of an LZ or lane usually impacts both the current operation and other operations. When the use of either LZ will not affect the scheme of maneuver or plan of supporting fire of adjacent units, the HUC, in coordination with the AMC or AFL, may be delegated the authority to use the alternate zone to exploit a tactical advantage or to improve the ground situation. If the use of a selected zone will affect adjacent or higher units, this authority cannot be delegated below the highest unit affected.

Helicopterborne Forces

Organization

The organization of forces may include some or all of the elements of the MAGTF.

Note

Throughout the remainder of this publication, the term HTF will include the ground combat element (GCE), aviation combat element (ACE), combat service support element (CSSE), and command element forces used to conduct helicopterborne assault.

Since task organization is essential in the conduct of helicopterborne operations, the helicopterborne force, as a part of the MAGTF, is an integrated force tailored to a specific mission under the command of a single commander. Typically, The MAGTF commander directs the formation of an HTF and designates a mission commander. The AMC, escort flight leader (EFL), AFL, and HUC are subordinate to the mission commander. To exploit opportunities offered by an HTF, commanders and leaders must understand the principles upon which the HTF was organized and its intent of employment.

Task Organization Considerations

Planners must consider the following during task organization of the helicopterborne force:

- Availability and allocation of aviation assets.
- Task organization is determined and announced early in the planning process; and it should be included in the warning order.
- The helicopterborne force provides sufficient combat power to seize initial objectives and protect LZs.
- The helicopterborne force requires a mission-specific balance of mobility, combat power, and sustaining power.
- The required combat power is delivered to the objective area as soon as possible, consistent with aircraft and LZ capabilities, to provide surprise and shock effect.
- Ability of the helicopterborne force to arrive intact at the LZ (providing en route security, throughout the entire flight route, and during actions on the objective) and to facilitate follow-on operations.
- Development of an effective C2 system.
- Combat support elements are normally placed in direct support to the helicopterborne force to ensure close coordination and continuous, dedicated support throughout an operation.

Capabilities

A helicopterborne force provides commanders with unique capabilities. No other ground force on the battlefield can respond to a tactical situation and move considerable distances as quickly as a helicopterborne force. It can extend the battlefield, move with great agility, and rapidly concentrate combat power. Specifically, helicopterborne forces can—

- Attack enemy positions from any direction.
- Overfly or bypass barriers/obstacles and strike objectives on otherwise inaccessible areas.
- Conduct deep attacks and raids beyond the forward line of own troops (FLOT) or point of contact by using helicopters to insert and extract forces.
- Rapidly concentrate, disperse, or redeploy to extend the area of influence.
- Provide responsive reserves thereby allowing commanders to commit a larger portion of their forces to action.
- React rapidly to tactical opportunities and necessities.
- Conduct exploitation and pursuit operations.
- Rapidly place forces at tactically decisive points in the battle area.
- Provide surveillance or screen over a wide area.
- React to rear area threats.
- Rapidly secure and defend key terrain such as crossing sites, road junctions, or bridges.
- Bypass enemy positions.
- Achieve surprise.
- Conduct operations under adverse weather conditions and at night to facilitate deception and surprise.
- Conduct fast-paced operations over extended distances.
- Conduct economy of force operations over a wide area.
- Rapidly reinforce/sustain committed units.

Limitations

A helicopterborne force is light and mobile, and it relies on helicopter support throughout the operation. As such, it may be limited by—

- Severe weather and winds.
- Extreme temperatures.
- Reliance on air lines of communications.
- Enemy aircraft, air defense, and electronic warfare (EW) action.
- Reduced ground mobility once inserted.
- Availability of suitable LZs and PZs.
- Available nuclear, biological, and chemical (NBC) protection and decontamination capability.
- Reduced vehicle-mounted antitank weapon systems.
- Battlefield obscuration.
- High fuel and ammunition consumption rates.
- Limited accessibility to supporting arms, especially indirect fires.

Another critical helicopterborne operation's planning consideration is the inherent trade off between the range a helicopter can fly and the fuel, troops, and cargo it can carry. Therefore, the following factors impact distance, fuel, and carrying abilities:

- Weather conditions at the PZ and LZ determine the maximum weight (i.e., fuel, troops, cargo) a helicopter can safely lift.
- The fuel load determines the distance the helicopter can fly.
- The distance of the mission determines the fuel required.
- The remaining excess payload after fuel determines how many troops and cargo can be moved the desired distance.
- Movement of a desired number of troops and cargo determines the remaining payload available for fuel.

- The use of en route refueling via forward arming and refueling point (FARP) or aerial refueling can minimize the trade off between range and payload.

Vulnerabilities

Helicopterborne forces use the helicopter to move and close with the enemy. Initial assault elements must be light and mobile; therefore, they are often separated from heavy weapon systems, supporting arms, equipment, and materiel that provide protection and survivability on the battlefield. A helicopterborne force is particularly vulnerable to—

- Attack by enemy air defense weapon systems during the movement phase.
- Attack by NBC systems, because of limited NBC protection and decontamination.
- Attacks (ground, air, or artillery) during the loading and unloading phases and at other times when infantry is not dug in.
- Electronic attack due to heavy reliance on radio communications for command and control.

CHAPTER 2

PREPARATION FOR COMBAT

Helicopterborne forces must prepare for combat operations by following troop leading procedures and organizing for a specific mission. This chapter discusses procedures and organizations that provide a basis for detailed discussion of helicopterborne operations in later chapters. The following paragraphs discuss combat preparation procedures for helicopterborne operations:

- Intelligence preparation of the battlespace (IPB).
- Threat.
- Task organization.

Intelligence Preparation of the Battlespace

IPB's main purpose is to support commanders and their staffs in the decisionmaking process. It integrates enemy doctrine with the weather and terrain, the mission, and the specific battlefield environment in order to produce a graphic intelligence estimate that portrays probable enemy courses of action. Because of aircraft vulnerability during helicopterborne operations, IPB's systematic approach to analyzing the enemy, weather, and terrain in a specific geographic area makes it critical to helicopterborne operations. The four steps of IPB are—

- Define the battlespace environment.
- Describe the battlespace's effects.
- Evaluate the threat.
- Determine threat courses of action.

Once hostilities begin and current information becomes available, the IPB estimate becomes dynamic, changing with the immediate situation on the battlefield. See Marine Corps Reference Publication (MCRP) 2-12A, *Intelligence Preparation of the Battlespace*, for a detailed discussion of this process.

Threat

Threat capabilities vary based on the enemy and the situation. However, there are basic threats to helicopterborne operations that will not change:

- Air defense fires, including small arms fires, must be identified and addressed by effective suppressive measures and increased emphasis on accurate and timely enemy intelligence.
- Fixed-wing and rotary-wing aircraft capabilities and limitations within the area of operations must be understood and measures taken to minimize the risk of encounter.
- EW capabilities, to include jamming, direction finding and monitoring of communications/radars, must be considered and appropriate countermeasures employed.
- Threat actions to counter PZ/LZ operations (analyze threat capabilities that could interdict friendly PZs/LZs with ground forces, indirect fires, and aerial attack).

Planners and commanders must constantly evaluate the threat in terms of the forms of contact available to the threat force. Forms of contact include observation, indirect fire, direct fire (to include air defense fires), obstacles, NBC, air, reserve forces, and EW. Countering the threat requires knowledge of enemy doctrine, tactics, and equipment and the capability to find and exploit enemy weak points with helicopterborne forces. IPB provides commanders and planners with an analytical methodology that reduces uncertainty concerning the enemy, the environment, and the terrain in order to determine and exploit enemy weaknesses.

Task Organization

Task-organized HTFs conduct helicopterborne operations. This task organization involves organizing both ground and aviation assets and requires coordination, planning, and execution between both ground commanders and aviation commanders to execute the ground tactical plan.

Note
Normally there is only one ground element in a helicopterborne operation although units may be landed in different locations.

The force is structured around an infantry unit and can vary in size from a reinforced rifle company to a reinforced Marine regiment. The HTF is normally part of a MAGTF and designed to accomplish a specific mission.

The MAGTF commander normally directs the formation of an HTF and allocates dedicated air resources. The MAGTF commander designates and assigns the mission commander. The mission commander allocates assets and defines authority and responsibility by designating command and support relationships. The mission commander ensures that ground operations are conducted according to the commander's intent and assists the commander in integrating the helicopterborne operation into the overall operational plan. Combat support and CSS are task-organized to provide the full range of support necessary to accomplish the helicopterborne mission. Commanders supporting the helicopterborne operation ensure that support operations are conducted according to the needs of the total helicopterborne force, to include both ground and supporting aviation units.

Aviation support is task organized to fully support all facets of the helicopterborne operation. Aviation support includes all or a portion of the six functions of Marine aviation in varying degrees, based on the tactical situation and the helicopterborne force's mission. The designated AMC ensures that all supporting operations are executed in a manner that best supports the ground tactical plan.

An HTF exists until completion of a specific mission. The MAGTF commander or designated mission commander establishes the criteria that constitutes mission completion under which the aviation elements can return to their parent unit.

Table 2-1 identifies specific organization, roles, and missions based on task organization.

Table 2-1. Task Organization of Helicopterborne Operations.

Asset	Organization	Role	Mission
Infantry	Units typically form nucleus of HTF. Units range from reinforced company through reinforced regiment. Units must prepare to assume helicopterborne missions.	Operate under the control of the HUC in direct support role to the HTF.	As directed by the HTF mission commander.
Assault Support Helicopters	One or more reinforced helicopter squadrons support the HTF.	Operate under the control of the AMC in a direct support role to the HTF.	Combat assault transport providing tactical mobility for troops, equipment, and weapon systems by internal and external load. Aerial resupply by internal and external load. Recovery and evacuation of equipment. Casualty evacuation or other air evacuation. Dedicated or hasty tactical recovery of aircraft and personnel.
Attack Helicopters	Task organized.	Operate under the control of the AMC in a direct support role to the HTF.	Fire support against point targets and/or antiarmor operations (e.g., air interdiction, close air support). Armed escort for assault support operations. SEAD artillery and other weapons en route to and during insertions and/or extractions. Observation of the LZ and objective areas to neutralize enemy resistance and to block enemy attempts to reinforce the objective area. Escort for tactical recovery of aircraft and personnel (TRAP) forces and/or security for downed aircraft. Armed and visual reconnaissance. Fire support and coordination and terminal control for supporting arms forward air controller (airborne) (FAC[A]).

Table 2-1. Task Organization of Helicopterborne Operations.

Asset	Organization	Role	Mission
Utility Helicopters	Task organized.	Operate under the control of the AMC in a direct support role to the HTF.	Enhance command, control, and communications capability for the HTF. Fire support against point targets (close air support). Armed escort for assault support operations. Observation of the LZ and objective areas to neutralize enemy resistance and to block enemy attempts to reinforce the objective area. Rescue escort for TRAP and security for downed aircraft. Armed and visual reconnaissance. Fire support coordination and terminal control for supporting arms (FAC[A]).
Unmanned Aerial Vehicles	Task organized.	Operate under the control of the AMC in a direct support role to the HTF.	Reconnaissance of PZs, flight routes, LZs, and objectives. Forward observation (or all around) of ground forces to provide limited early warning.
Artillery Fire Support	Normally, an artillery battalion provides support; however, artillery batteries may be required.	Operate under the control of the mission commander in a direct support or attached role to the HTF. Provide rapid response capability to prepared LZs and objectives. Provide suppression of enemy artillery and air defense fires.	Disruption of threat artillery/indirect fires. SEAD along flight routes and in the vicinity of LZs. LZ preparation. Screening fires. Deception fires. Artillery raids/aerial repositioning. Delivery of family of scatterable mines. Objective preparation and/or suppression.
Air Defense	Task-organized low altitude air defense (LAAD) assets.	Operate under the control of the mission commander in coordination with the AMC in a direct support or attached role to the HTF. Equipped with light, air-transportable, short-range, man-portable air defense systems in order to fly with the lead assault elements and provide protection in the objective area.	Air defense of high value locations including PZ, LZ, objective areas, helicopter FARPs, and holding sites. Direct fires for ground defense.

Table 2-1. Task Organization of Helicopterborne Operations.

Asset	Organization	Role	Mission
Engineers	Units range from platoon through company.	Operate in a direct support role to the HTF. Typically, engineers are attached to infantry units during unit movement, but revert to general support once communications with parent HQs are re-established. Organize to move with the infantry and provide mobility, countermobility, and survivability construction equipment. Provide light engineering support if heavy-lift helicopters are available.	Construction and improvement of PZs and LZs. Construction of expedient counter-mobility obstacles using natural materials and demolitions. Construction of firing positions. Clearance of obstacles/minefields. Emplacement of minefields. Conduct of assault and covert breaches. Combat.
Electronic Warfare	Task organized.	Provide EW planning and operations support to the HTF.	Electronic attack. Disruption of enemy command, control, and communications. Degradation of enemy fire support and air defense radio nets. EW support. Collection of electronic intelligence.
Reserve Elements	Task organized.	Provide reinforcement or assumption of another unit's mission.	As required.
CSSE	Task organized.	Provide mission-specific support to the HTF throughout the operation.	As directed by the HTF mission commander.

CHAPTER 3

COMMAND AND CONTROL

Command and control is the process of directing and controlling the activities of a military force in order to obtain an objective. Since the battlespace over which the helicopterborne force operates may extend beyond the typical battlespace of a company- through regiment-sized force, operational command and control must be given special considerations.

A helicopterborne C2 system must communicate orders, coordinate support, and provide direction to the helicopterborne force in spite of great distances, enemy interference, and the potential loss of key facilities and individuals. Above all, this system must function quickly and effectively, thus allowing the helicopterborne force to receive and process information and to make decisions faster than the enemy. An effective helicopterborne C2 system includes the procedures, facilities, equipment, and personnel required to gather information, make plans, communicate changes, and control all ground and air elements in pursuit of the objective.

Planning

The MAGTF commander addresses C2 requirements and establishes an effective C2 system early in the planning phase. An effective helicopterborne C2 system allows the MAGTF commander to direct diverse, widely dispersed air and ground elements between the initial PZ and the final objective. Since helicopterborne operations are subject to degraded communications due to the extended distance from which they operate, the MAGTF commander must develop a C2 plan and system that allow execution of the mission despite degraded radio communications. The key to successful helicopterborne command and control lies in effective task organization, precise planning, decentralized execution, and the use of helicopterborne radio nets. See chapter 4 for detailed planning information.

Effective Task Organization

All assets must be tailored into discrete, task-organized elements, each with two-way radio communications, unity of command, clearly defined missions and objectives, and provisions for maintaining unit integrity throughout the operation. An effective task organization, with each element having a clearly defined mission, allows the HTF the flexibility to decentralize execution and ensures mission success despite degraded communications, the fog of battle, or unexpected enemy reaction.

Precise Planning

Helicopterborne operations must be precisely planned and well briefed before execution so that each subordinate leader knows exactly what is expected, knows the commander's intent, and knows that the mission can be executed despite the loss of radio communications. Contingencies or alternatives must be built into each plan to allow for continuation of the mission in a fluid operational environment.

Typically, precise planning is done through the use of time driven or event driven actions. Time driven actions occur at specific times. Event driven actions occur relative to each other. For example, a time driven action is the firing of an LZ artillery preparation precisely from H-5 minutes to H-1 minutes. If previously planned, this can be executed with degraded communications.

An example of an event driven action is the insertion of one company into the alternate LZ if the lead company makes enemy contact on the primary LZ. If previously planned, this event occurs as expected and without the need for lengthy radio communications.

Decentralized Execution

Although it is centrally planned, the execution of a helicopterborne operation is decentralized. Subordinate commanders are given the maximum possible freedom of action (consistent with safety and mission accomplishment considerations) to ensure mission accomplishment.

Helicopterborne Radio Nets

Radio nets that facilitate ground-to-ground, air-to-air, and ground-to-air communications are established to provide for the timely flow of information and redundancy in capability. This helps reduce, if not eliminate, the loss or degradation of communications.

Airborne Command and Control

The ability to place command and control in the air allows the mission commander to personally influence the operation, communicate with subordinates, and arrive at a timely decision. The mission commander and his staff are positioned where they can best support the mission and, sometimes, this may be airborne in a C2 platform with the AMC or in an assault support helicopter with the AFL. In a large helicopterborne operation with multiple LZs, subordinate commanders may also require C2 helicopters to control and coordinate their units.

Typically, the mission commander commands airborne only during air movement and the initial stages of the landing. When a major portion of the assault elements have landed, the mission commander displaces to a forward command post on the ground. The mission commander should avoid routinely controlling ground operations from the air. This can lead to over supervision of subordinate units and can sometimes give an inaccurate picture of the true tactical situation. Appendix B summarizes the essential items included in the planning phase of helicopterborne operations.

Mission Commander

The MAGTF commander allocates assets, defines both authority and responsibility by designating command and support relationships, and designates the time that the HTF is established. The MAGTF commander may be the mission commander depending on the scope of the helicopterborne operation or may designate a mission commander. The MAGTF commander designates a mission commander as follows:

- If a regiment HTF is designated, the regimental commander is the mission commander.
- If a battalion HTF is designated, the battalion commander is the mission commander.
- If a company HTF is designated, the battalion commander is the mission commander.

The mission commander exercises command via the established command and support relationships, is responsible for the planning and execution of all aspects of the assigned mission, and determines when the HTF is disbanded.

For a regiment HTF, the mission commander may designate the commanding officer of the main effort as the HUC or may designate each of the battalion commanders as subordinate mission commanders. This provides a standard, yet flexible, C2 architecture that is scalable and meets the requirements for any mission or contingency for which an HTF might be employed. The role of the mission commander is to ensure a unity of command throughout the operation.

Air Mission Commander

The commander of the aviation unit tasked to support the helicopterborne operation designates the AMC. The AMC is the Marine aviator designated by the commander of the aviation unit tasked to support a helicopterborne operation. Depending on the size and scope of the MAGTF, he may also be the ACE commander. Unless the mission commander is the MAGTF commander, there will not be a command relationship between the mission commander and the AMC. In some cases, the mission commander exercises tactical control of assigned aviation assets; that is, he may direct and control the movements or maneuvers necessary to accomplish missions or tasks assigned. During the planning phase, the AMC is co-equal to the HUC. During execution, specific authority is delegated from the mission commander to the AMC. The AMC typically works in direct support of the mission commander and answers directly to the mission commander's requests for assistance and support. The supported-supporting relationships and the means by which they are executed are critical to mission success; therefore, the AMC must have a detailed understanding of the command and support relationships with key subordinates (e.g., AFL, EFL). The AMC is responsible for the planning and execution of all aviation functions relative to the assigned helicopterborne mission; therefore he must be an experienced aviator. It is the AMC's responsibility to establish liaison with the mission commander and HUC (the commander responsible for the ground tactical plan) in order to conduct concurrent and parallel planning. The AMC shall assume the duties of the assault support coordinator (airborne) (ASC[A]) of a mission if no ASC(A) is assigned.

Helicopterborne Unit Commander

The HUC is a ground officer who has been designated as the commander of the helicopterborne force:

- For a regiment HTF, the HUC is either the main effort battalion commander or any battalion commander as otherwise stipulated.
- For a battalion HTF, the HUC is either the commanding officer of the main effort company or any company commander as otherwise stipulated.
- For a company HTF, the company commander is the HUC.

As such, the HUC is charged with executing and accomplishing the ground tactical plan and with coordinating aviation and the other support required to plan and execute the helicopterborne mission. The HUC's unit composes the helicopter landing force. Normally, there is only one HUC commanding a single helicopterborne unit although there may be multiple lifts and landings. As in any operation, the HUC must move in order to see the battlefield and where he can control the operation. Depending on the situation, the HUC can be airborne during the movement and insertion phases. At other times, the HUC fights the battle from a tactical command post deployed well forward. The HUC is subordinate to the mission commander and co-equal to the AMC during the planning phase. During execution, specific authority is delegated from the mission commander to the HUC.

Assault Flight Leader

The AFL is an experienced aviator in command of the assault support flight. The AFL reports to the AMC and assists in the planning of flight routes, LZs, and all other facets of the helicopterborne mission that directly involve assault support aircraft. The AFL is subordinate to the AMC and is co-equal to the EFL during the planning phase. During execution, specific authority is delegated from the AMC to the AFL. (See app. C for a checklist to assist with planning.)

Escort Flight Leader

The EFL is an experienced aviator in command of the escort flight. The EFL reports to the AMC and assists in the planning of LZ preparation, fire support planning, threat mitigation, and all other facets of the helicopterborne mission that directly involve attack aircraft. The EFL is subordinate to the AMC and is co-equal to the AFL during the planning phase. During execution, specific authority is delegated from the AMC to the EFL.

Assault Support Coordinator (Airborne)

The ASC(A), an experienced aviator operating from an aircraft, is delegated the authority to perform specific coordination and control functions of helicopter operations and to provide situational awareness to the helicopterborne force during a specific evolution. Typically, the ASC(A) provides information concerning—

- Weather along the approach and retirement routes and in the LZs.
- Observed enemy operations that may affect the HTF mission.
- Changes to helicopter routes.
- Changes in the friendly situation.
- Employment of supporting arms, to include tactical air coordinator (airborne) (TAC[A]) activities.

The ASC(A) may directly support a mission commander or be employed as an extension of the direct air support center (DASC) or helicopter direction center to coordinate assault support activities that do not warrant the assignment of a mission commander. If employed as an extension of the DASC/helicopter direction center, these agencies assign specific functions to the ASC(A) (e.g., initial assaults, subsequent assaults).

The ASC(A) is also responsible for coordinating the activities of all helicopters in his assigned area. If employed in conjunction with the TAC(A) or FAC(A), and no mission commander is assigned, the relationship with the ASC(A) is established by the tactical air commander or his designated representative. When an ASC(A) has not been designated, the AMC discharges the duties of the ASC(A) within the limits of his authority. To facilitate timely and coordinated decisions affecting helicopterborne assaults, the ASC(A) and a representative of the HUC should be assigned to the same aircraft if feasible.

Tactical Air Coordinator (Airborne)

The TAC(A) is an extension of the DASC and coordinates with the ground commander's tactical air control party (TACP); subordinate FAC(A); and the mortars, artillery, and naval gunfire shore fire control parties. Normally, the TAC(A) is the senior coordinator having the authority over all aircraft operation within his assigned area.

Forward Air Controller (Airborne)

The FAC(A) is an aviator and forward air controller (FAC) who is airborne (in either a helicopter or fixed-wing aircraft) in the area of operations. The FAC(A) is an extension of the TACP and his primary function is the detection and destruction of enemy targets through close air support (CAS) and deep air support. The FAC(A) is assigned as either direct support of a ground unit or as a subordinate to the TAC(A) or ASC(A) that provides air control as required. The

FAC(A) performs the following tasks within his assigned area of responsibility:

- Detects enemy targets for suppression, neutralization, and destruction.
- Controls CAS missions.
- Performs strike coordination and armed reconnaissance when directed.
- Controls LZ preparations.
- Marks targets and LZs.
- Controls mortar, artillery, and naval gunfire missions when required.
- Conducts visual reconnaissance.
- Reports intelligence information to the appropriate ground or air control agency.

Initial Terminal Guidance Teams

Initial terminal guidance (ITG) teams from the force reconnaissance company or reconnaissance battalion, Marine division, have the inherent capability to provide terminal guidance for the initial helicopter waves in the LZs. However, all ground units must be trained, equipped, and capable of performing ITG for small helicopter landings. ITG teams are comprised of personnel who are inserted into LZs in advance of the landing zone control team, and they may be the first elements to make contact with the enemy. Therefore, they must promptly report any enemy activity that may influence the landing. ITG teams execute prelanding reconnaissance tasks and establish and operate signal devices that guide the initial helicopter waves from the initial point to the LZ. The use of ITG teams may increase the difficulty or even prevent the use of LZ preparation fires due to the presence of friendly troops in or around the LZ. Duties of the team may include—

- Determining obstructions in the LZs, including radiological hazards.
- Giving advance notice of enemy positions.
- Selecting PZs/LZs.
- Marking LZs for day and night.
- Recommending use of alternate LZs.
- Controlling supporting arms.
- Recommending actions to be taken by following waves.
- Organizing an area around the zone to stage troops, equipment, or supplies to be picked up or moved upon landing.
- Selecting an initial point near the LZ.
- Establishing communications with approaching flight.
- Giving an LZ brief (see app. D) to the flight leader.

Helicopter Support Team

The helicopter support team (HST) is a task organization whose composition is formed and equipped for employment in PZs and LZs. These teams facilitate the pickup, movement, and landing of helicopterborne troops, equipment and supplies, and the evacuation of selected casualties and prisoners of war (POWs). The team usually includes a headquarters element, a helicopter control element, and an LZ platoon. The LZ platoon provides supply and engineer support functions. The helicopter control element consists of a landing zone control team provided by the ACE commander when necessary and may include personnel to provide refueling and emergency maintenance. The landing zone control team may be task-organized from the Marine air traffic control detachment when the size or scope of the operation warrants and the MAGTF commander determines it is necessary. The Marine air traffic control detachment tasks may include—

- Installing and operating air traffic control and navigational systems required for the control of aircraft at expeditionary airfields and remote landing sites.
- Providing air traffic control services that facilitate the safe, orderly, expeditious flow of aircraft within designated terminal/landing areas.
- Maintaining the capability to deploy independent air traffic control teams/units.

Pickup Zone Control Officer

Typically, the HUC designates a pickup zone control officer (PZCO) (either a FAC or air officer) from the supported unit of each PZ. The PZCO organizes, controls, and coordinates operations in the PZ and pushes elements out of the PZ. He operates on a designated tactical net and is prepared to assist in executing needed changes. The PZCO is the key individual during night operations or when multiple subordinate elements are being lifted from the same PZ.

Subordinate Unit Commanders

Subordinate unit commanders are attached to the HUC and normally function as they would in any other infantry task force. Each subordinate unit commander must be prepared to receive other elements for movement.

Command Post

The command post provides command and control for the execution of helicopterborne operations. It must be mobile and well forward. It is normally helicopter lifted into the objective area soon after the initial echelon. A C2 helicopter may serve as a command post if enemy air defense systems allow.

Combat Operations Center

The combat operations center is normally established in the command post. It provides planning for future operations and ongoing operations as directed by the commanding officer. Functions of the combat operations center include—

- Monitoring current operations and maintaining current enemy and friendly situations.
- Gathering and disseminating intelligence.
- Keeping higher and adjacent organizations informed of the friendly situation.
- Submitting recurring reports.
- Providing liaison to higher and adjacent organizations.
- Establishing a fire support coordination center (FSCC).
- Coordinating combat support, aviation, engineer, and air defense.
- Advising the commander on the use of combat support for current and future operations.
- Monitoring airspace and coordinating supporting fires.
- Continuing planning for future operations and overseeing the preparation of all contingency plans.
- Issuing combat/warning orders as necessary.

Rear Area Operations

The rear area provides uninterrupted support to the force as a whole. Both operational level and tactical level logistic operations occur within the rear area. Typically, the HTF is not assigned a rear area responsibility. Rather, it stages and launches from the rear area of its higher headquarters. A CSSE assigned to the HTF will likely collocate with the logistic trains of this headquarters to facilitate the coordination of support to the HTF.

CHAPTER 4

PLANNING

During the planning phase of the helicopterborne operation, it is essential that coordination among the mission commander, HUC, and AMC begin as soon as possible. While the HUC and AMC must plan the operation together, the mission commander's and HUC's ground concept of operations drives all planning for the helicopterborne operation. The battalion is the lowest level that has sufficient personnel to plan, coordinate, and control a helicopterborne operation. When company-sized operations are conducted, the bulk of the planning and air-ground asset coordination takes place at the battalion headquarters.

Helicopterborne operations require the development of five basic plans: ground tactical plan, landing plan, air movement plan, loading plan, and staging plan. The HUC directs the formulation of the ground tactical plan, the landing plan, and the loading plan. The AFL is principally responsible for formulating the air movement plan. During planning of a helicopterborne operation, one primary consideration for the HUC and AMC is the enemy air defense situation. Other planning considerations include, but are not limited to, sortie rates and aircraft types, availability, and capabilities.

To achieve the necessary, rapid buildup of combat power, a helicopterborne operation requires the massing of helicopters. As a planning figure, a minimum of one third of the ground unit must be landed in the zone in the first wave, but should always be based on a detailed assessment of the threat within the objective area.

The basis for planning the timing of the operation is L-hour-the time when the first assault helicopter in the first wave touches down in the LZ.

SECTION I. ESTIMATION PROCESS

A vital portion of planning is the estimation process. The following are actions that are important parts of the estimation process. Some of these actions are required to initiate the estimation process, some actions are initiated to keep the estimation process in a continual, progressive state based on the evolution of the operation, and some actions are options for the commander to implement based on the battle situation.

- The helicopterborne force uses command and staff actions and troop leading procedures common to other combat operations to execute planning.
- Planning of helicopterborne operations is as detailed as time permits, and, if time allows, complete written orders and plans are developed. However, if a tactical opportunity does not allow detailed planning, then rapid planning is used and the planning steps are compressed or conducted concurrently and detailed written plans and orders are replaced with standing operating procedures (SOPs) or lessons learned. See appendix E for a sample battalion SOP.
- Information flow is critical to the successful completion of a helicopterborne operation. Information is received from higher headquarters and all echelons provide information intended to reduce the planning burden of subordinate units.

- All tactical estimates used in troop leading procedures employ the factors of mission, enemy, terrain and weather, troops and support available–time available (METT-T). Analysis of METT-T provides data that is used during the estimation process to reach a decision.

Analysis of METT-T

The analysis of METT-T is an important part of the estimation process because it formulates the design of the commander's plan of attack and contributes significantly to the estimate of the tactical situation for helicopterborne operations. Appendix F provides a guide for detailed, precise reverse planning.

Mission

Mission analysis is conducted early in the estimation process. The mission includes the critical tasks that must be accomplished. The tasks are either specified tasks stated by the order or implied tasks determined by the commander. Mission analysis determines not only what must be accomplished, but also the intent of the commander ordering the mission. It also states the restraints and constraints placed on the mission by the higher headquarters. This analysis provides the basis for task organization and must be conducted to determine if it is more advantageous to strike with a helicopterborne force or attack with a ground force. To analyze the mission, the following questions must be asked:

- Does the mission require the rapid massing or shifting of combat power over an extended distance?
- Does the mission require surprise?
- Does the mission require the flexibility, mobility, and speed afforded by helicopters?
- Since helicopterborne operations are inherently a high risk operation that can yield a high payoff, does the payoff warrant the risk?
- What is the level of training? See appendix G.

Enemy

The evaluation of the enemy defines the enemy's capabilities and most probable courses of action. The following factors about the enemy must be considered:

- Identification: Who is he?
- Location: Where is he? Where is he going?
- Disposition: How is he organized? What are his formations?
- Strength: What are his strengths versus friendly forces strengths?
- Morale: What are his esprit, experience, state of training, and regular or reserve forces?
- Capabilities: What are the obstacles, indirect fires, direct fires, observation, NBC, air, reserve forces, and EW?
- Composition: What is his armor, infantry (motorized or light), artillery, and combat support?
- Probable courses of action: What is his likely mission or objective? How will he probably achieve it? (Think about the most probable course of action and most dangerous course of action.)

When planning a helicopterborne operation, the following factors about the enemy must be considered:

- Enemy air defense weapons and capabilities.
- Enemy mobility; particularly the ability to influence potential flight routes and helicopter LZs.
- Enemy NBC capability; particularly the ability to react to the insertion.
- Capability to interdict or interrupt helicopter movements with enemy helicopters or fixed-wing aircraft.
- Enemy EW capability.

Terrain and Weather

In all military operations, terrain analysis is conducted by the criteria described in the acronym OCOKA-W [observation and fields of fire, cover and concealment, obstacles, key terrain, avenues of approach, weather]. In helicopterborne operations, these factors must be analyzed in relation to their effects on the force during movement to the PZ, loading, air movement, LZ insertion, movement to the objective, and subsequent actions.

Observation and Fields of Fire

The following considerations relate to both enemy and friendly forces:

- Enemy visual observation and/or electronic surveillance of PZs, flight routes, and LZs.
- Enhanced friendly observation provided by aerial and ground reconnaissance assets.
- Ease of navigation along flight routes particularly for night or adverse weather operations.
- Ability to influence the PZs, LZs, and flight routes with indirect and direct fire.

Cover and Concealment

The following considerations relate to the cover and concealment of friendly forces:

- Terrain masking for low level flight routes and insertions.
- Covered battle positions for attack helicopters.
- PZs and LZs that offer ground forces cover and concealment.

Obstacles

While helicopters can bypass most obstacles, PZs and LZs must be free of natural and/or manmade obstacles that could preclude a helicopter landing or affect the ground scheme of maneuver.

Key Terrain

The possession of key terrain provides a decided advantage to a force and, in many instances, is mission-dependent. However, in helicopterborne operations, key terrain is not limited to that which influences the ground scheme of maneuver. It must also be analyzed in terms of—

- PZs and LZs.
- Flight routes.
- Attack helicopter battle positions.
- Occupation of enemy positions, especially enemy air defense assets.
- Potential FARPs.

Avenues of Approach

Air and ground avenues of approach are considered in both offensive and defensive operations from both friendly and enemy viewpoints. A good avenue of approach for a helicopterborne force offers–

- A reasonable degree of mobility and few if any natural or manmade obstacles to the aircraft.
- Little or no canalization.
- Terrain masking that decreases effectiveness of enemy air defense weapons.
- Cover.
- Concealment.
- Good lines of communication and logistics.
- Ease of link up with other forces when appropriate.

Weather

Weather can greatly impact an operation. Changes in the weather may result in an interruption of helicopter support and require changes in planned operations. Considerations include–

- Fog, low clouds, heavy rain, and other factors that limit visibility for pilots.
- Illumination and moon angle during night vision device (NVD) operations.
- Ice, sleet, and freezing rain may cause ice accumulation on airframes, which can become catastrophic.
- High temperatures and/or density altitudes that degrade aircraft engine performance and lift capability.
- Darkness, normally an advantage to well-trained pilots and ground forces.
- High winds (large gust spreads).
- Weather conditions that create hazards on PZs and LZs, such as blowing dust, sand, or snow.

Troops and Support Available

Troops and support available encompasses not only troops to be lifted into the objective LZ but also all combat power to include combat support and CSS available to the helicopterborne force. The helicopterborne force should have enough combat power relative to the threat to seize initial objectives and protect the LZs until follow-on echelons arrive in the objective area, and helicopter lift capability is the single most important variable in determining how combat power can be introduced into the objective area.

Aircrew endurance must also be considered. For planning purposes, the AFL usually considers 8 hours of flight time within a 24-hour period for dual-piloted aircraft to be a safe limit for aircrews. Day and night operations may require different day endurance considerations for the crew. If those limits are exceeded during a single period, then degraded aircrew performance can be expected.

Time Available

The time available to prepare for a helicopterborne operation is extremely important. It is often the scarcest resource and is vital to planning. The commander must adjust the planning process to make optimum use of this perishable resource. When time is critical, the commander's intuition, judgment, and experience are invaluable in guiding his staff and subordinate commanders. This planning must be coordinated between the GCE and the ACE. While the ACE is preparing its aircrews, servicing its aircraft, and planning its flights, the GCE is also preparing for its mission. Once the GCE's and ACE's concurrent planning is completed, these elements must be brought together for rehearsals, especially if an unusual mission is being planned. The following must be considered:

- Allocating the time required to prepare, plan, and rehearse. Helicopterborne planning must be centralized and precise. It normally requires more time than other operations.
- Additional planning time may be required for night operations and those involving multiple PZs and/or multiple LZs.
- The HUC must allow adequate time to ensure that all subordinates and support elements are thoroughly briefed. Briefing time can be significantly reduced with SOPs and previous training.

Ground Commander's View of the Plan of Attack

From the ground commander's point of view, the plan of attack for a helicopterborne operation includes the scheme of maneuver, plan for supporting fires, loading plan, air movement, and the landing plan, which are developed concurrently and are closely integrated.

Scheme of Maneuver

The scheme of maneuver is the tactical plan executed by a force in order to accomplish its assigned mission. It includes objectives, LZs, forms of maneuver to be employed, distribution of forces, and necessary control measures. The commander must take into account the time required for the helicopterborne force to consolidate at the LZ and move to its objective. The ability to secure the LZ and move to the objective must be compared to the enemy's ability to reinforce the threatened area.

Plan for Supporting Fires

The plan for supporting fires is the commander's employment of all supporting arms to assist in the accomplishment of the mission. The plan for supporting fire supports the scheme of maneuver and provides for SEAD during the helicopter approach and retirement, preparation of the LZ, fires in support of the consolidation of the LZ, and fires in support of the ground operation. The plan for supporting fires should also include plans for employment of air defense weapons.

Loading Plan

The loading plan is designed to establish, organize, and control activities in the PZ, plan for the movement of troops and equipment to the PZ, and establish the priority of loading units. For battalions or larger, a written plan may be required. However, the requirement for written loading instructions can be minimized by advanced planning and detailed unit SOPs. Regardless of its simplicity, the loading plan must receive command attention during planning. The ultimate success of the operation is directly related to a properly developed loading plan and subsequent control of unit loading.

Air Movement Plan

The air movement plan provides for the control and protection of the helicopterborne force during the air movement. The air movement plan is primarily the responsibility of the AMC, although the ground commander also contributes to its development. The air movement plan includes the selection of approach and retirement lanes, control points, and en route SEAD and the provisions for escort by attack helicopter or other aviation.

Landing Plan

The landing plan consists of the commander's guidance concerning the desired time, place, and sequence of arrival of units. The landing plan must support the ground tactical plan.

SECTION II. DETAILED PLANNING

The HUC and AMC receive a mission from the mission commander. If possible, the HUC and AMC should receive the mission together in order to begin initial coordination and to facilitate concurrent/parallel, detailed planning. The procedures for detailed planning are as follows:

- The HUC receives a mission from the mission commander.
- The HUC and AMC pass mission information to their staffs. They may do this in the form of a warning order to allow their staff and subordinates to begin general planning.
- It is the mission commander's responsibility to make initial liaison with the HUC and AMC. During his initial liaison, the mission commander gives the HUC and his staff planning data relative to the numbers and types of helicopters available for the lift.

During detailed planning, there are five basic plans that comprise the reverse planning sequence for each helicopterborne operation. The five basic plans are the ground tactical plan, landing plan, air movement plan, loading plan, and staging plan. These plans should not be developed independently. They are coordinated and developed concurrently by the staff of the designated helicopterborne force.

Ground Tactical Plan

The foundation for a successful helicopterborne operation is the HUC's ground tactical plan. Normally, the ground tactical plan is developed first and is the basis from which the other plans are derived. The ground tactical plan is a portion of the helicopterborne scheme of maneuver. The ground tactical plan specifies actions in the objective area that ultimately accomplishes the mission. The plan also includes subsequent operations that can include link-up operations, repositioning of the force, and sustainment. The ground tactical plan for helicopterborne operations contains essentially the same considerations as any other infantry form of maneuver except that it must capitalize on surprise, speed, and mobility in order to achieve mission success.

Mission

The most obvious portion of the helicopterborne mission is the requirement to conduct a helicopter movement. The following concerns should be examined:

- Missions of all task force elements and methods of employment.
- Purpose of the helicopter movement.
- Reason for using helicopters.
- Task organization of ground, air, and combat support units.
- Number of Marines to be lifted.
- Total weight to be lifted.
- Internal and external loading.
- Approximate distance of the air movement.

Concept of Operations

The ground concept of operations is formulated in five parts:

- Ground movement to the PZ.
- Securing and organizing the PZ (include the task organization of the force and embarkation plan for force aboard mission aircraft).
- Actions at and securing of the LZ.
- Ground movement from the LZ to the objective or actions on the objective if the force is landed directly on the objective.
- Operations subsequent to securing the objective.

Landing Plan

The landing plan must support the ground tactical plan. The plan sequences troops and equipment into the area of operations so that units arrive at locations and times prepared to execute the ground tactical plan. Appendix F provides more information. Considerations in developing the landing plan include—

- Availability, location, size, and enemy proximity to potential LZs are overriding factors.
- The helicopterborne force is most vulnerable during landing.
- Troops and equipment must land with tactical integrity.
- To avoid disorientation, troops must be informed if changes occur in the landing directions that were given during their initial brief.
- Initially, there may be no other friendly units in the area; therefore, the helicopterborne force must land prepared to fight in any direction.
- The landing plan should offer flexibility so that a variety of options are available in developing a scheme of maneuver.
- Supporting fires (i.e., artillery, naval surface fire support, CAS) must be planned in and around each objective area LZ.
- Although the objective may be beyond the range of supporting artillery fire, artillery or mortars may be brought into the LZ early to provide fire support for subsequent lifts on the objective.
- The plan includes provisions for TRAP, immediate re-embarkation, emergency extract, resupply, and casualty evacuation by air.

Selection of Landing Zones

LZs are selected during planning between the HUC and the AFL in coordination with the intelligence officer (S-2). The selected LZs are approved by the MAGTF commander or mission commander. Regardless of whether the site is a LZ or PZ—

- The ground chosen must support the safe landing of helicopters.
- The selected site should be identifiable from the air.

- The enemy situation must be such that the site can be secured without undue interference from enemy fires.

Each helicopter requires a different size LZ/PZ, and each area needs to be on level ground. Lighting conditions also affect the size of the LZ for each helicopter: daylight zones should be 100 feet larger than the diameter of aircraft rotor blades and night zones should be 150 feet larger than the diameter of aircraft rotor blades. Table 4-1 provides the recommend landing zone diameters for different types of helicopters.

Table 4-1. Recommended Landing Zone Diameters.

Type Aircraft	Rotor Blade Diameter (feet)	Landing Zone Diameter Daylight (+100)/Night (+150)
UH-1	50	150/200
CH-46	85	185/235
CH-53	100	200/250

Ideally, each LZ/PZ is as level as possible, and free of major obstacles that might obstruct landings or takeoffs (e.g., tall trees, telephone/power lines). Plans must be made to mark or identify obstacles than cannot be removed in order to aid the aircrew's ability to safely land the aircraft. The ground itself must be firm enough to prevent bogging down, otherwise aircraft may have to hover during loading or unloading operations. The site must be free of heavy dust, loose snow, logs, rocks, or dry grass.

Although level ground is preferable, some areas that can support the helicopter will not be level. As a planning rule, LZs with slopes greater than 7 degrees require additional consideration by the AFL and AMC.

LZ/PZ entry and exit routes are chosen to ensure that takeoffs or landings can occur over the lowest obstacles and that the direction is into the wind with minimum crosswinds of 10 knots and tailwinds of no more than 5 knots. Wind direction must also be considered in terms of its effect on the dust created by the helicopter's landing and takeoff.

The helicopter must be able to ascend or descend vertically into the LZ/PZ when fully loaded. The landing point for each helicopter should be at a distance 10 times as far from an obstacle as the obstacle is high (see fig. 4-1).

LZs are selected using the following criteria:

- Ground commander's concept of operations.
- LZs can be located on, near, or away from the objective, depending on the factors of METT-T.
- The size determines how much combat power can be landed at one time. This also determines the need for additional LZs or separation between waves.
- An alternate LZ should be planned for each primary LZ selected to ensure flexibility.

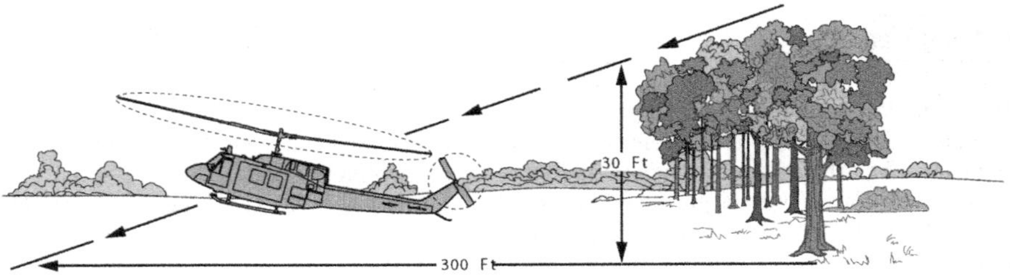

Figure 4-1. LZ Obstacle Clearance.

- Enemy troop concentration and air defenses and their capability to react to a nearby helicopterborne landing are considered when selecting LZs.
- LZs deny enemy observation and acquisition of friendly ground and air elements while they are en route to, in, and departing from the LZ.
- If possible, the helicopterborne force should land on the enemy side of obstacles when attacking and use obstacles to protect LZs from the enemy at the other times.
- LZs must be free of obstacles.
- LZs should be readily identifiable from the air. When possible, reconnaissance units should be used to reconnoiter and mark the LZ.
- Requirements for logistic support.
- Requirements for fire support.
- Available lanes to and from LZs and any restrictive effects on the employment of supporting arms.
- Reduced visibility or strong winds may preclude or limit the use of marginal LZs.

Single Versus Multiple Landing Zones

In addition to deciding where to land in relation to the objective, planners must address whether to use single or multiple LZs. The advantages of a single LZ are as follows:

- Allows concentration of combat power in one location (if the LZ is large enough).
- Facilitates control of the operation.
- Concentrates supporting arms in and around the LZ. Firepower is diffused if more than one LZ preparation is required.
- Provides better security during subsequent lifts.
- Requires fewer attack helicopters for security.
- Reduces the number of flight routes in the objective area, making it more difficult for enemy intelligence sources to detect helicopterborne operations.
- Centralizes any required resupply operations.
- Concentrates efforts of limited LZ control personnel and engineers on one LZ.
- Requires less planning and rehearsal time.

The advantages of multiple LZs are as follows:

- Avoids grouping assets in one location and creates a lucrative target for enemy mortars, artillery, and CAS.
- Allows rapid dispersal of ground elements to accomplish tasks in separate areas.
- Reduces the enemy's ability to detect and react to the initial lift.
- Forces the enemy to fight in more than one direction.
- Reduces the possibility of troop congestion in one LZ.
- Eliminates aircraft congestion in one LZ.
- Makes it difficult for the enemy to determine the size of the helicopterborne force and the exact location of supporting weapons.

Landing Formations

Aircraft formations on the LZ should facilitate operational offloading and deployment. The number and type of aircraft and the configuration and size of the LZ may dictate the landing formation during the planning process. Because contact is expected in the LZ, troops are landed ready to employ fire and movement. In order to reduce troop exposure, an LZ formation will not maintain standard distances between aircraft and must land rapidly in a safe area as close to concealment as possible.

Supporting Fires

Typically, the initial assault is made without preparatory fires in order to achieve tactical surprise. However, preparatory fires are planned for each LZ so they can be fired if needed. Planned fires for helicopterborne operations should be intense and short, but with a high volume of fire to maximize surprise and shock. Supporting fires end just before the first assault element's landing. Fire support plans developed to support the landing plan must address the following:

- False preparations fired into areas other than the objective or LZ.
- Loss of surprise due to lengthy preparations.
- Time, location, speed, and size of committed forces that affect the ability to achieve surprise.
- Based on the allocation of fire support and the commander's guidance, the fire support coordinator (FSC) conducts fire support planning to support the landing. Higher echelons allocate supporting arms to lower echelons. Commanders at lower echelons may then further allocate fire support (e.g., priority of fires).
- Significant targets and either the known or suspected enemy, regardless of size, warrants target planning.
- Obstacles to landing and maneuver; for example, some ordnance used to prepare the site can cause craters, tree blowdown, fires, and LZ obscuration and therefore may not be desirable.
- Fires are scheduled to be lifted or shifted to coincide with the arrival times of aircraft formations.
- Positive control measures must be established for lifting or shifting fires. Airspace coordination areas may be necessary to protect approach and retirement lanes.
- Use of escorts as CAS for the GCE in the LZ must be coordinated prior to mission launch. The HUC should not assume that escort aircraft will be available if their use has not been coordinated with the AMC.
- If a FAC(A) is available, consideration should be given to the FAC(A) controlling fires during insertion of the initial wave(s) until the GCE can assume control of fires.

Emergency Extract

If the enemy force has overwhelming combat power, the supported unit must plan for an emergency extract. The supported unit must address what the helicopterborne force expects the ACE to provide. Support may involve a surge of CAS, the maneuver of another reinforcing unit into a nearby LZ, or it may be emergency extraction. Units must understand, however, that an emergency extract may not always be possible. If the unit has suffered casualties or the enemy force possesses greater mobility, an attempt to execute an emergency extract may be disastrous.

The HUC may be delegated the authority by the mission commander to call for an emergency extract if an enemy force threatens to destroy the unit. Therefore, the HUC must clearly understand how the helicopterborne mission fits into the MAGTF scheme of maneuver. The repercussions of executing an emergency extract on the MAGTF scheme of maneuver must weigh heavily on the decision to execute.

Once the decision has been made to execute an emergency extract, escort and/or combat air patrol aircraft move into position to provide suppressive fires. The helicopterborne unit breaks contact and moves to a secure PZ. On call, the assault support aircraft extracts the unit. Consideration should be given to landing all aircraft in one wave—zone, threat, and aircraft availability permitting—in order to expedite the extract.

Since confusion is inherent during an emergency extract, all participants must thoroughly understand weapons conditions in order to reduce/eliminate the risk of fratricide. A visual means to distinguish between friend and foe must be established and briefed to the assault support gunners.

Air Movement Plan

The air movement plan is based on the ground tactical plan and the landing plan. The air movement plan specifies the air movement schedule and provides instructions for the air movement of troops, equipment, and supplies from PZs to LZs. It also provides coordinating instructions pertaining to approach and retirement routes, air control points, aircraft speeds, altitude, and formations. The planned use of attack helicopters, to include security and linkup locations, should be included in the plan. During MAGTF operations, the MAGTF commander approves the air movement plan based on the recommendations of the AMC, HUC, and the CSS unit commander. The information essential to the ground combat commander is obtained and disseminated by the ground unit air officer.

WARNING

In order to prevent fratricide, it is imperative that the fire support plan for the entire helicopterborne operation be thoroughly coordinated between the mission commander, HUC, and AMC. The AMC has primary responsibility for planning/coordinating fires in support of the air movement plan. The HUC has primary responsibility for planning/coordinating fires for the remainder of the helicopterborne operation. Fires in support of the landing plan must be particularly well coordinated between the HUC and AMC. If an EFL is assigned, he will be the lead ACE fire support planner during the development of the fire support plan for the entire helicopterborne operation.

Selection of Helicopter Approach and Retirement Routes

Helicopter approach and retirement routes are air corridors in which helicopters fly to and from their destination during helicopter operations. During MAGTF operations, the MAGTF commander, based on the recommendations of the AMC and the HUC, selects helicopter approach and retirement routes. Examination of METT-T with strong emphasis on threat analysis determines approach and retirement route selection. Route and altitude are interdependent in the selection and are considered concurrently to determine the optimum movement. Avoidance of enemy detection and fires is the primary consideration. In addition to METT-T factors, route selection planning considers the following general principles:

- Has the terrain been used to the best tactical advantage?
- Have the primary and alternate approach and retirement routes been identified?
- Have routes been selected that are easy to identify and navigate?
- Will communications capabilities be impaired?
- Can the routes be used under adverse weather conditions?
- Do unique support requirements exist for the routes selected?
- Have supporting arms capabilities and limitations been considered?

Ground Considerations During Helicopter Movement

Ground commanders must know the dimensions of helicopter approach and retirement routes for fire support planning, and they must be familiar with where routes begin and end. Because helicopters may be forced to land along the route, commanders should be able to identify prominent terrain features and checkpoints during flight in order to facilitate mission accomplishment. Therefore, ground commanders should conduct visual reconnaissance to the maximum extent possible during flight and, as helicopters pass over checkpoints, inform Marines so they can prepare to disembark.

Supporting Fires Along the Approach and Retirement Route

The following must be considered during planning of supporting fires for helicopter approach and retirement routes:

- Fires along the flight route are planned to suppress known or suspected enemy positions. These fires should be of short duration. Multiple target engagement techniques are used (e.g., groups, series).
- Fire plans cover the PZs, flight routes, and LZs. Fire support plans include SEAD systems and smoke to protect formations from enemy detection. This requires aggressive fire planning by the FSC and direct coordination with supporting units.
- All available fire support is used to suppress and/or destroy enemy weapons.
- Support may consist of smoke or other electronic attack for suppressing or confusing enemy air defense systems. However, smoke can become an obscurant that can interfere with the flight phase of the operation.
- On-call fires are planned along the flight route to ensure the rapid adjustment of targets of opportunity.
- During night operations, the use of illumination fire requires detailed planning because illumination can interfere with NVDs and cause unsafe conditions.

Loading Plan

Mobility is accomplished only to the extent that the ground unit retains its ability to accomplish its primary mission while moving. If the helicopterborne unit must re-organize or adjust upon landing, the mobility, momentum, tempo, and initiative are lost or diminished. The only way to maintain the required degree of mobility during a helicopterborne operation is to load and land helicopters in the manner and sequence that allows immediate assumption of the mission upon landing. Therefore, an effective and efficient loading plan is critical to the success of a helicopterborne operation. Appendix H provides helicopter characteristics that will assist in load planning.

The loading plan is based on the air movement plan. It ensures that Marines, equipment, and supplies are loaded on the correct aircraft. Helicopter loads are also prioritized to establish a bump plan. A bump plan ensures that essential Marines and equipment are loaded ahead of less critical loads in case of aircraft breakdown or other problems. In any case, planning must cover the organization and operation of the PZ including load positions, day and night markings, and communications. The loading plan is more critical when mixing internal and external loads and/or when mixing helicopter types.

Primary and Alternate Pickup Zones

Identification of primary and alternate PZs is the first step in developing the loading plan. The establishment of adequate unit SOPs covering PZ operations and loading plans reduces the requirement for detailed, written plans. PZs must accommodate helicopter landing and lift, be securable with ground forces, and facilitate staging of aircraft loads. Although enemy contact or influence is not desired during pickup, pickup under pressure must be considered and addressed during planning. Ideal PZs also facilitate delivery of suppressive fires, security for ground troops, and security for landed and lifting helicopters.

Pickup Zone Criteria

PZ criteria for selection and organization are similar to those required for an LZ. However, additional PZ criteria are as follows:

- Number: multiple PZs avoid concentrating forces in one area.
- Size: each PZ accommodates all supporting aircraft at once, if possible.
- Proximity to troops: if possible, PZs should not require ground movement to the PZ by troops.

- Accessibility: if possible, PZs should be accessible by vehicles to move support assets and infantry.
- Vulnerability to attack: selected PZs should be masked by terrain from enemy observation to the maximum extent possible.
- Preparation: if possible, PZs are usable as-is rather than requiring clearing

Once PZs are identified, the ground commander selects and assigns PZs to be used by subordinate units.

Pickup Zone Control

PZs are organized to meet specific mission requirements. Depending on the size and magnitude of the operation, a PZ may be as small as one point in one site or it may include numerous sites.

A multiple site PZ may require select sites to conduct strictly internal or external loading functions. Specializing sites for specific functions (e.g., heavy lift, external operations, combat assault transport) may facilitate operations when employing a PZ that contains numerous sites within its confines. For example, a battalion-sized helicopterborne assault could require a four-site zone: one site may conduct cargo external lift functions, another site external vehicle lifts, another internal cargo loading, and another tactical loading of troops. This enables equipment and personnel to be concentrated where most needed.

Pickup Zone Control Officer

The PZCO organizes, controls, and coordinates operations in the PZ. The PZCO is designated by and responsible to the commanding officer of the moving unit. In battalion-level operations, the PZCO could be the battalion executive officer. In most situations, the air officer will not function as the PZCO. He will usually be in the combat operations center assisting the FSC in the allocation of air power. Marines from the service platoon, headquarters and service company, form the PZ control group and are trained by the battalion air officer. All personnel allocated to PZ control must be completely trained in HST functions to include external lifting terminal control. The air officer uses HST Marines from the landing support battalion to augment the training program. The PZCO accomplishes the following:

- Forms a control group to conduct operations and to provide assistance. The control group may include—
  - Terminal control.
  - Guides to lead elements from unit positions around the zone to the staging area.
  - Marines to conduct hookup operations for external lifts.
  - Marines to clear PZs and to provide local security.
  - Other Marines as needed to perform required tasks within the zone.

Note

For battalion helicopterborne operations, each company commander may need to appoint a PZCO to operate a company PZ for the battalion.

- Establishes communications on two primary radio frequencies: one to control movement and loading of units and one to control aviation elements. Alternate frequencies are provided as necessary.
- Plans and initiates fire support near PZs to provide all-round protection (from available support) without endangering arrival and departure of troops or aircraft. The fire support plan must be closely coordinated with the HUC and AMC.
- Plans and initiates adequate security to protect the main body as it assembles, moves to the PZ, and is lifted out. If the PZ is within the friendly area, other forces, if available, provide the security elements. Security comes from the helicopterborne force's resources if the force is to be extracted from the objective area.
- Clears the PZ of obstacles.
- Marks the PZ.

PZ Identification and Marking

Identification and marking occurs as follows:

- The PZCO directs the markings of PZs, sites, and points.
- PZs are designated by a code name.
- A zone with more than one site is usually identifiable in the air by prominent terrain features.
- A zone can be marked by several methods (e.g., colored smoke, air panels). Red is never used to mark an aircraft landing position. Red is used to mark landing obstacles (e.g., trees, stumps) in the landing area.
- Landing sites are designated by a color.
- Landing points are designated by two-digit numbers.
- Regardless of the type of markers, the PZ is marked to indicate where aircraft are to land and to coincide with the selected PZ aircraft formation.
- An effective method is to have several individuals in each unit paint (and carry) an extra camouflage cover or a modified (cut to size) air panel. The colored cover, when displayed, indicates where the lead aircraft lands.

Movement to the PZ

Ground and air unit movement to the PZ is scheduled so that only the troops and the helicopter to be loaded arrive at the PZ at the same time. This prevents congestion, preserves security, and reduces vulnerability to enemy actions on the PZ. To coordinate the movement of units to the PZ, the PZCO—

- Selects troop assembly areas, holding areas, and routes of movement. A holding area is located close to the PZ. It is used only when the assembly is some distance away and does not allow timely movement to the PZ.
- Determines the movement time of ground units to PZs.
- Specifies arrival time(s) and sees that movement of units remains on schedule.

Helicopter Wave and Serial Assignment Table

At company and lower levels, the helicopter wave and serial assignment table (HWSAT) assigns each Marine and major items of equipment or supplies to a specific aircraft. The HWSAT is a simple accountability tool that provides a loading manifest for each aircraft. If time is limited, the HWSAT can be a simple list of each Marine (by name) and the equipment to be loaded on each aircraft and given to a specified representative. Either method of accountability ensures that if an aircraft is lost, a list of on board personnel and/or equipment is available.

Note

If recurring, small-unit helicopterborne operations are anticipated, small-unit leaders may require their Marines to carry individual, preprinted 3- by 5-inch cards for quick collection upon loading.

One of the critical datum calculated during HWSAT preparation is the gross weight (personnel and equipment) to be loaded on the aircraft. The pilot in command of each aircraft must be provided with the gross weight of each load to ensure that aircraft weight limitations are not exceeded and that the aircraft can safely accomplish its assigned mission with the proposed load (given the ambient environmental conditions).

Load Planning

During preparation of the loading tables, all unit leaders attempt to maintain the following:

- Fire teams and squads are loaded intact on the same aircraft and platoons in the same wave in order to maintain the tactical integrity of each unit.
- The composite first wave is an exception to maintaining tactical integrity of units. The composite first wave facilitates the seizure of the objective LZ and the landing of subsequent waves into the LZ area. Establishing a composite first wave enables the commander

to task-organize the initial landing with varying elements from subordinate units of the tactical force. This option may facilitate securing the objective LZ by eliminating the need to move troops on the ground as subsequent waves land.

- Each unit load should be functionally self-sufficient whenever possible.
- Every towed item is accompanied by its prime mover.
- Crews are loaded with their vehicle(s) or weapon(s).
- Component parts are loaded with major items of equipment.
- Ammunition is carried with the weapon.
- Sufficient personnel are on board to unload cargo.
- Communications between flights is established.
- Tactical spread loading is applied to all loads so that all leaders, or all crew-served weapons, are not loaded on the same aircraft. Thus, if an aircraft is lost, the mission is not seriously hampered. For example, loading the platoon commander, platoon sergeant, and all squad leaders on the same helicopter or loading more than one machine gun team on the same aircraft are violations of cross-loading principles.

Another consideration is to determine whether internal or external loading is the best delivery method for equipment and supplies. Helicopters loaded internally can fly faster and are more maneuverable. Externally (sling) loaded helicopters fly slower and are less maneuverable; however, they can be loaded and unloaded more rapidly than internally loaded helicopters. Externally loaded supplies can also present problems if supplies are destined for more than one location or unit. The loading method used depends largely on availability of sling and rigging equipment.

Aircraft Bump Plan

AMCs must inform the PZCO about any changes to the number, type, and carrying capability of the aircraft en route to the PZ. The PZCO must have time to re-organize sticks and institute the bump plan before the arrival of the assault support aircraft. Each aircraft load has a bump sequence designated on its helicopter employment and assault landing table (HEALT). Bump priority ensures that the most essential personnel and equipment arrive at the objective area first.

Note

A load (stick) is the smallest group of personnel and/or equipment that will be moved by a single aircraft and will not be broken into smaller units. More than one stick may be moved by a single aircraft (i.e., 2 sticks of 12 may embark one CH-53E).

The bump plan specifies personnel and equipment that may be bumped and delivered later. If all personnel within the load cannot be lifted, individuals must know who is to offload and in what sequence. This ensures that key personnel are not bumped arbitrarily. Also, a bump sequence is designated for aircraft within each serial or flight. This ensures that key aircraft loads are not left in the PZ. When an aircraft within a serial or flight cannot lift off and key personnel are on board, they offload and reboard another aircraft that has priority.

Company or larger units specify a PZ bump-and-straggler collection point. Personnel not moved as planned report to this location, are accounted for, regrouped, and rescheduled by the PZCO for later delivery to appropriate LZs.

The mission to be accomplished by each subordinate unit upon landing determines the sequence of departure from each PZ. Unit priorities are based on the sequence of arrival at their LZs. For example, if company A is to land first (at L-hour), and company B second (at L+15), and company B is 15 minutes farther (in flight time) from the objective LZ; it may depart the PZ before company A.

Staging Plan

The staging plan is based on the loading plan and prescribes the arrival time of ground units (troops, equipment, and supplies) at the PZ and their proper order for movement. Loads must be ready before aircraft arrive at the PZ; usually, ground units are expected to be in the PZ 15 minutes before aircraft arrival.

The staging plan also restates the PZ organization, defines flight routes to the PZ, and provides instructions for linkup of all aviation assets. Air-to-air linkup of aviation units should be avoided, especially at night when NVDs are being used.

SECTION III. MISSION BRIEFING AND DEBRIEFING

The responsibility for operational briefings is a function of command and rests with the commander tasked with executing the helicopterborne operation. Generally, each subordinate level of command conducts a briefing that focuses on that unit's participation in the operation. For example, a command representative of the landing force/MAGTF briefs the overall operation, representatives from the helicopter and helicopterborne units brief their unit's participation, the AMC briefs the airborne portion of the helicopterborne assault, each individual flight leader briefs his flight, and each individual aircraft commander briefs his aircrew.

Mission Brief

The mission brief is the final phase of the planning effort and should be attended by all key personnel. This brief sets forth the concept of operations, scheme of maneuver, and specific details concerning mission coordination and execution. Information is provided that enables each participant to understand the overall operation and his specific role and responsibilities regarding mission execution. Joint briefings with representatives from each participating unit should be used as much as possible. Depending on the mission requirements, the minimum attendees should include the AMC, HUC, FSC or his representative, TAC(A) or FAC(A), FAC, fixed-wing attack aircraft flight leader, attack helicopter flight leader/flight coordinator, AFL, and the aircraft commanders. The information developed during the planning effort becomes the subject matter for the mission brief.

Helicopterborne Mission Briefing Guide

Proper briefing of flight crews is essential to mission success. The helicopterborne mission briefing should be conducted in the most logical, brief, and organized manner possible. The information from appendix F and the helicopter tactical manual series of publications (Naval Warfare Publication [NWP] 3-22.5) provides detailed mission briefing guides. The following information is presented at the helicopter mission briefing:

- Helicopter assignment.
- Call signs.
- Flight leader/alternate flight leader.
- Execution timeline.
- Controlling agencies.
- Frequencies.
- Radio procedures.
- Identification, friend or foe procedures/codes.
- Navigation data.
- HEALT and HWSAT.
- Execution checklist.
- LZ/PZ diagrams (include imagery if available).

It must be emphasized that the mission briefing guide is only that, a guide, and is not intended to be used in total, or depicted sequence, for every type of the mission. Only those items directly applicable to a specific mission should be incorporated into the mission brief. Since all members of the mission will not be involved in the planning, it is imperative that the flight brief be well delivered, organized, and easily understood. The use of the mission briefing guide, adequate rehearsals, and the use of tactical SOPs significantly reduce the time required to conduct the brief.

CHAPTER 5

COMBAT OPERATIONS

Helicopterborne operations are deliberate, precisely planned, and vigorously executed combat operations. Helicopterborne operations are designed to allow friendly forces to strike over extended distances and terrain barriers in order to attack the enemy when and where he is most vulnerable. It is the MAGTF commander's option to employ a helicopterborne operation to enhance a ground operation. The decision to conduct a helicopterborne operation depends on many factors relative to METT-T. The commander uses the HTF when the situation permits and when the possible payoff outweighs the risk. A helicopterborne capability promotes speed, surprise, and flexibility so that the commander can react rapidly to a changing situation.

A helicopterborne operation can be conducted alone or in conjunction with another operation. A helicopterborne operation is based on the ground tactical plan, and it capitalizes on speed and flexibility in order to gain maximum surprise. The ultimate goal of a helicopterborne operation is to place the assault echelon on or near the objective capable of immediate action. Typically, helicopterborne operations are conducted–

- In offensive operations to–
  - Seize key terrain.
  - Overcome obstacles.
  - Conduct raids.
- Insert or extract patrols–
  - Conduct security operations.
  - Support deception operations.
  - Reposition forces.
- Rapidly reinforce a successful attack.
- In defensive operations to–
  - Block enemy penetrations or withdrawals.
  - Reinforce encircled forces.
- Insert or extract patrols–
  - Conduct security operations.
  - Conduct counterattacks.
  - Reposition forces.

SECTION I. HELICOPTERBORNE OPERATIONS IN OFFENSIVE OPERATIONS

The helicopterborne attack is the basic type of offensive operation conducted by the helicopterborne task force. It is the integration of combat, combat support, and CSS elements into and out of an objective area. Generally, the term insertion applies when discussing the movement of an assault into the objective area and the extraction applies when discussing the movement from the objective area. These terms are fundamental to all helicopterborne operations.

Attack

The opportunity to attack may arise during the course of battle or it may be created by skillful, tactical leadership. Whatever the source, the attack should be fast, violent, and coordinated. The helicopterborne force may conduct an attack in conjunction with other forces. The type of action conducted by the larger force usually dictates the type of attack employed by the helicopterborne force.

There are two types of attacks that a helicopterborne operation may be involved in: hasty and deliberate. The hasty attack is one in which preparation time is traded for speed in order to exploit an opportunity. A deliberate attack is characterized by preplanned, coordinated employment of firepower and maneuver. The major difference between a hasty and deliberate attack is the time and enemy information available.

Hasty Attack

Because the hasty attack is conducted on short notice, there is little time to plan and orders must be brief. The helicopterborne force must then rely on previous training and SOPs to cover these situations. The following may require a helicopterborne force to execute a hasty attack in support of a larger force:

- Unexpected contact is made during the movement to contact by the larger force (the helicopterborne force can be used to exploit a tactical advantage or further develop the tactical situation).
- When part of the larger force's deliberate attack plan is modified while the operation is underway (the helicopterborne force can reinforce a weakened area or exploit a tactical advantage).
- An exploitation and/or pursuit (the helicopterborne force is committed to exploit the attack's success and maintain momentum).
- A counterattack.

When a hasty attack is considered under any of the circumstances listed above, it is important to identify tentative PZs, LZs, and approach and retirement lanes throughout the higher unit's zone of action. Continuously identifying these areas permits rapid commitment of the helicopterborne force anywhere in the sector.

When the helicopterborne force is committed, the commander initiates several actions simultaneously. The commander directs suppressive fires to neutralize the enemy's ability to counter the helicopterborne operation and concentrates sufficient combat power to overwhelm the enemy at selected points. While the helicopterborne force is en route, supporting fires suppress or destroy known or suspected enemy positions. Priority of fires go to SEAD.

When the attack starts, attack helicopters overwatch and react as necessary while the FSC and air officer direct artillery, mortars, CAS, and other supporting fires. Artillery and mortars destroy, neutralize, or suppress enemy indirect fire weapons as soon as they are located. Smoke may be used to screen aircraft movement from observation. However, the commander is careful that smoke does not obscure the LZ and hinder the landing operation. Airspace coordination must be effected early.

Deliberate Attack

The helicopterborne force, as part of a larger force operation, may conduct a deliberate attack. The helicopterborne force is provided sufficient time to develop a detailed, coordinated plan; receive additional assets; change task organization as necessary; and gather detailed intelligence. Detailed information about the terrain is collected so that the best PZs, LZs, and flight routes can be selected. During a deliberate attack, helicopterborne objectives are normally in the enemy's rear area or the attack is from the flank or rear.

When the larger force concentrates its combat power on a narrow front to break through the enemy defense, the helicopterborne force may bypass main defenses to destroy artillery positions, command posts, logistic and communications facilities, and/or to secure key terrain in the enemy's rear.

An attack against a heavier or well-prepared enemy force, particularly on the mechanized and/or armor battlefield, may subject the helicopterborne force to devastating firepower. For this reason, the helicopterborne force may land away from the objective and conduct an infantry attack in conjunction with friendly mechanized and/or armor forces. The helicopterborne force must also consider that a highly mobile enemy force could encircle the force before it moves from an LZ. Consequently, the commander selects LZs in terrain that is favorable to infantry operations and employs antitank weapons and attack helicopters on likely armor avenues of approach. With accurate intelligence, these actions provide time to organize after landing and to attack the objective.

Exploitation

Exploitation is another operation where the helicopterborne operations can be used for a low-risk, high-payoff operation. Exploitation is an operation undertaken to follow up success in the attack. Attacks are conducted with two overriding requirements: speed and violence. The attackers bypass pockets of resistance to concentrate on the destruction of the more vulnerable headquarters, combat support, and CSS units. They disrupt the enemy's command and control; his flow of fuel, ammunition, and repair parts; and his air defenses and artillery. Disrupting the enemy's flow of support weakens and/or destroys the enemy. Enemy air defenses are avoided or suppressed so the helicopterborne force can exploit the situation.

Pursuit

Pursuit is an offensive operation designed to catch or cut off a hostile force that is attempting to escape, with the intent of destroying it. Its purpose is to envelop the retreating force and destroy it by coordinated fire and maneuver. A helicopterborne force, operating as part of the pursuit force, can expect to be ordered to bypass resistance of any kind and move relentlessly to deep objectives that serve as chokepoints for the retreating enemy. The helicopter provides the task force with the high degree of mobility required to conduct pursuit operations.

Fixed-wing aircraft, attack helicopters, and the helicopterborne force can repeatedly attack the flanks of the withdrawing enemy columns, slowing them and aiding in their destruction. Blocking positions can be established on withdrawal routes to trap enemy forces between the encircling force and the direct pressure force. Artillery and FARPs should be lifted into the encircling force areas as soon as possible.

Secure and Defend

Seizure and Retention

A helicopterborne assault operation conducted during a seizure and retention operation is two-phased and requires detailed planning like a deliberate attack. The secure and defend mission is conducted when an objective, such as a vital terrain feature, must be seized and retained. The limited staying power of the helicopterborne force dictates early link up with ground units, reinforcement by other units, or extraction from the enemy area.

First Phase

The first phase is an attack to secure terrain to be controlled by the helicopterborne force in the initial stages of the assault. This should be a single lift insertion of sufficient combat power to defeat enemy forces at the objective.

Second Phase

The second phase of the operation is defense of the objective LZ. This normally involves a perimeter defense that controls all terrain essential to the defense of the LZ. This area should be large enough to provide operating space for combat, combat support, and CSS units. It may require adequate PZs for simultaneous helicopterborne operations and space for landing artillery, follow-on forces, and supplies. The area must be small enough for the helicopterborne unit to defend, yet large enough to permit defense-in-depth and maneuver of Reserves to counter enemy attacks. Size is dictated by mission, enemy strength, enemy disposition, terrain, and helicopterborne combat power.

Control Measures

Boundaries delineate responsibilities of the helicopterborne force's subordinate elements. A battalion area may be divided into company-sized zones and objectives. Each company clears, secures, and defends an assigned area. The size sector assigned each company should be within its capabilities to seize and defend. For example, a company facing a dangerous avenue of approach is assigned a smaller sector than a company facing a less dangerous avenue. Defensive responsibilities for an avenue of approach are not divided. The unit assigned the approach also covers any dominating terrain.

Assault Objectives

If a terrain feature is vital to mission accomplishment and is to be secured during an assault, it is designated as an assault objective. The assault objective should include terrain that dominates all high-speed approaches to the objective area. Assault objectives are assigned priorities. Those specified by higher headquarters are given first priority. Others are ranked according to the threat they would pose if controlled by the enemy. A company's sector should include at least one LZ for an assault landing, resupply, and evacuation.

Reconnaissance in Force

Reconnaissance in force is an offensive operation designed to discover and/or test the enemy's strength or to obtain other information. Typically, it is conducted when the enemy situation is vague. The information (e.g., major weaknesses in enemy positions) obtained from the operation, if promptly exploited, may provide a significant tactical advantage. The reconnaissance in force is planned and conducted with units specifically prepared to find the enemy and to develop the situation. The reconnaissance in force accepts risk in order to gain intelligence information rapidly and in more detail than other reconnaissance methods. The MAGTF commander assigning a helicopterborne force a reconnaissance in force mission must determine the following:

- Is the desired information important enough to justify the risks to personnel and possible loss of aircraft?
- Can other intelligence methods obtain the same information in sufficient time with less risk?
- Will a reconnaissance in force compromise future plans?
- Can the operation succeed?

A reconnaissance in force operation also locates the enemy and presses him into reacting. When a weak point is discovered, the helicopterborne force exploits it quickly. The commander exercises caution, however, since the enemy response may be too strong for the helicopterborne force. Thus, the commander also plans a withdrawal to avoid destruction.

However, a helicopterborne operation is not designed to land where the enemy is located. Therefore, the helicopterborne operation, in many cases, will be a reconnaissance in force.

The reconnaissance in force is an ideal mission for the helicopterborne force in an insurgent environment to keep constant pressure on a guerrilla force. Helicopterborne forces are also suited for reconnaissance in force operations against conventional light infantry. Helicopterborne forces are not suited for operations against strong armor threats due to the likelihood of ground contact with an enemy force that has superior firepower, mobility, and protection.

If the commander needs information about a particular area, a reconnaissance in force is planned and executed as an attack against a specific objective(s). The objective of the attack is usually of such importance that when threatened, the enemy will react. The helicopterborne force's combat power must be sufficient to force the enemy to react and reveal his position, strength, planned fires, and planned use of Reserves. It may also disrupt the enemy's planned operations and take the initiative from him. When a reconnaissance in force is used to achieve a specific objective, the helicopterborne battalion can deploy all its companies or the commander may commit one or two companies and retain the remaining elements to respond to tactical situations as they develop. When the enemy reacts to one unit, the units not in contact are shifted to exploit revealed enemy weaknesses or help extract a unit under pressure.

Raid

Helicopterborne forces are well-suited for raids; however, detailed planning and the elements of surprise are vital to the success of a helicopterborne raid. A helicopterborne raid is a swift penetration of hostile territory that may be conducted to destroy installations, confuse the enemy, or gather information, and it ends with a planned withdrawal. Because a raid is conducted behind enemy lines, it requires exact planning to ensure a high probability of success. The selection of LZs, PZs, and flight routes (as in the deliberate attack) is based on the results of detailed planning and required intelligence. Since the raiding force attempts to achieve surprise, the decision to land on the objective takes on added significance

SECTION II. HELICOPTERBORNE OPERATIONS IN DEFENSIVE OPERATIONS

Defense is a coordinated effort to deny the enemy his purpose in attacking. Types of defense include mobile and position. A mobile defense destroys the attacking enemy by offensive action. A position defense focuses on the retention of terrain. Based on the specific tactical mission, the commander uses METT-T analysis to determine the best type of defensive operation.

The helicopterborne force can defend against an infantry-heavy threat by employing its mobility to achieve a maneuver advantage over the enemy. This advantage can allow the force to perform operations in the security area, main battle area, and rear area.

The helicopterborne force may be able to conduct security force operations for a larger force. Normally, security area operations consist of air reconnaissance, infantry, artillery, engineer, and attack helicopter operations to position combat power and combat support quickly and in the most advantageous locations during rapidly changing situation. Infantry and artillery assigned to security force operations must be provided a security area that is organized based on the following:

- Number of enemy avenues of approach into the security area.
- Size and type of enemy forces.

Security area forces accomplish their mission by placing the majority of their combat units on the most dangerous avenues of approach into the security area. Air reconnaissance deploys to the front and provides early warning of the direction, speed, and composition of enemy forces. Enemy units are placed under fire as soon as they are within weapon range. As the enemy attempts to close with ground units of the security force, attack helicopters, artillery, and CAS provide firepower that enable ground units to displace by air to successive positions. Protection of helicopterborne infantry and antitank systems is achieved by superior mobility. Security area units wear the enemy down, deceive him as to the location of the main battle area, slow his speed of advance, cause him to mass, and may cause him to divulge his intentions. Units are assigned subsequent missions in the main battle area once the security area mission has been accomplished.

The main battle area is that portion of the battlefield in which the decisive battle is fought to defeat the enemy. It extends rearward from the forward edge of the battle area (FEBA) to the rear boundary of the command's subordinate units. The mobility advantage that the helicopterborne force has over the enemy's infantry-heavy units may allow it to defend in greater depth. The helicopterborne force defends by orienting on the destruction of advancing enemy forces and fights a series of battles in depth, attacking the enemy from the front, flanks, and rear while using minimal forces to maintain surveillance over the remainder of the assigned sector. Battle positions are selected and prepared throughout the main battle area along likely avenues of approach. Primary and alternate LZs and PZs are selected for each battle position. When enemy fires preclude extraction of the helicopterborne force from battle positions, covered and concealed routes are selected for ground movement to alternate PZs. Only when absolutely necessary should a helicopterborne force be directed to occupy or retain terrain. If there is a situation in which the retention of terrain is essential to the defense of the entire sector, its retention is specified.

Defensive Operations Against an Armor-Heavy Threat

The helicopterborne force is not well suited to defend against armored and mechanized forces. If it is used to defend against such forces, it should be employed in restrictive terrain not favorable to employment of massed armor. The HTF can be employed in built-up areas, mountainous terrain, and heavily forested areas. The helicopterborne force can conduct the following operations in the armored and mechanized battlefield in support of larger defensive operations:

- Main battle area operations in restrictive terrain.
- Economy of force or reserve.
- Rear operations.
- Flank security operations.
- Limited-objective counterattack operations or raids.
- Delay and withdrawal operations.
- Seizure of specific objectives for linkup operations.

Note

Attack helicopter and air reconnaissance units are best suited for employment with mechanized units.

Economy of Force

Defense in an economy of force role can be accomplished by displacing units of the helicopterborne force in depth on avenue of approach throughout the sector. The air reconnaissance elements can screen areas where enemy attack is possible but unlikely. Combat units are repositioned to counter the major enemy thrust. After engaging the enemy and before the enemy closes on battle positions, units are picked up from designated PZs and organized in depth. The helicopterborne force essentially conducts a delay. Field artillery is repositioned as necessary to halt the enemy advance. Attack helicopter elements should be placed in direct support of the ground commander. Elements of the helicopterborne force held in reserve are rapidly transported by helicopter into areas under enemy pressure.

Delay

The key to success in the delay is the commander's ability to array forces in depth before the initiation of the delay. Decisive engagement is accepted only to the degree and extent necessary to accomplish the delay mission. Contingency plans for stay-behind operations are developed. The helicopterborne force continually looks for and seizes the opportunity to launch small-scale, offensive helicopterborne assaults and attack helicopter raids into the enemy flanks and rear areas. A delay may be conducted to—

- Gain time so that other forces can deploy.
- Serve as an economy-of-force measure to allow concentration of friendly forces in other areas.
- Determine enemy composition, strength, intentions, and capabilities.
- Channel the enemy into selected areas and then destroy him.

The helicopterborne force should seldom be given a time-delay mission. This type of mission requires the force to delay the enemy for a specified time, restrict its mobility, and subject it to unacceptable losses. The delay in sector mission is more appropriate. The force disengages by helicopter before it is decisively engaged. Against armor forces, the force should displace at distances of no less than the maximum effective range of the enemy's most capable direct fire weapons (2,500 to 4,000 meters) and rely on attack helicopters to delay the armor while friendly infantry is extracted.

SECTION III. OTHER TACTICAL MISSIONS AND OPERATIONAL CONSIDERATIONS

Screening

A helicopterborne screening force provides early warning over an extended frontage. Screening missions are assigned to—

- Provide timely warning of enemy approach.
- Maintain visual contact and report movement.
- Engage enemy forces within capabilities.
- Avoid decisive engagement.

A screening mission employs a series of observation posts that overlook enemy avenues of approach and the areas between them. Patrols cover dead space between the observation posts, and they also cover other areas during limited visibility. When contact is established, the screening force withdraws on order, maintaining visual and/or electronic contact, and reports enemy movements. As in the delay, timely displacement is critical to force survival.

Guard Force

The helicopterborne force can perform flank or rear guard missions for a larger force and help protect the main body from ground observation, direct fire, and surprise attack. As a guard force, the helicopterborne force has sufficient combat power to attack enemy reconnaissance forces and to delay an enemy attack until the main body can deploy. The rear guard follows the main body, occupying successive positions. The rear guard also screens between flank positions and rear elements of the main body. The helicopterborne force can conduct rear guard operations by moving from position to position. These movements are controlled by using designated phaselines.

Covering Force

The air reconnaissance element can overfly rough terrain, find the enemy, and develop the situation. Units are deployed as necessary to ensure the uninterrupted movement of the main body. The covering force unit may use one of the following two methods to conduct the covering force mission:

- Reconnaissance units reconnoiter while the helicopterborne force remains in assembly areas or on order to be available for commitment. When contact is made with the enemy and after the air reconnaissance has developed the situation, the helicopterborne force is committed to destroy the enemy.
- Helicopterborne units conduct covering force operations as other forces move by bounds behind the leading unit. A helicopterborne force is normally not assigned as the covering force, but it may be assigned as a subordinate element of the covering force.

Reinforcement of Committed Units

An HTF can reinforce a committed unit:

- With uncommitted units (Reserves).
- With additional antitank assets.
- By moving artillery to weight the battle.

The MAGTF commander may also direct the insertion of a helicopterborne unit to reinforce threatened sectors and add depth to the battle area.

Antitank elements may be taken from a reserve unit or a unit that is not protecting an armor approach. Depending on the number of sections employed, the antitank element leader accompanies them for command and control. Careful consideration is given to planning the extraction of antitank elements because they may lack ground mobility in varying situations.

Artillery batteries can shift about the helicopterborne battlefield to ensure fire support to committed units.

Linkup Operations

When withdrawal of a helicopterborne force from the objective area is not planned or feasible, a linkup operation is conducted to join two forces. A helicopterborne force may participate as part of a larger force, or it may conduct a linkup with its own resources. Close coordination and detailed planning between the commanders of both units are essential.

River Crossing Operations

Helicopterborne forces may reduce CSS considerations during river crossing operations by—

- Flying over the river.
- Supporting bridge construction.
- Deploying reconnaissance units by air to verify and collect essential intelligence on crossing sites and enemy dispositions.
- Reaching objectives on the far shore quickly, eliminating enemy interference with development and use of crossing sites.
- Rapidly airlifting engineer bridging assets forward, eliminating traffic problems on the crossing site approaches.

If a deliberate crossing is chosen, the helicopterborne force, with its increased mobility, can be used to clear the near shore of enemy resistance. During the actual crossing, whether hasty or deliberate, the helicopterborne force can assist by—

- Attacking enemy forces that interfere with the crossing by seizing objectives that would be secure or assist in securing the bridgehead.
- Providing flank security.
- Securing crossing sites with or without smoke.

Rear Area Operations

Air reconnaissance can provide wide-area surveillance and security, and can be integrated into reaction force plans. Rear area operations are coordinated with the designated rear area commander. The helicopterborne force, as a potential reaction force, can be called upon to contain the enemy force if it does not have enough combat power to destroy it. Other forces would then be called upon to destroy the enemy.

The helicopterborne force also monitors likely infiltration routes and probable target areas for enemy attacks from the rear in order to counter enemy airmobile, airborne, or guerrilla infiltration threats. Probable LZs and PZs are identified and monitored by observation posts or remote sensors. Potential infiltration routes in unoccupied terrain are monitored with sensors to detect the enemy as early as possible.

Night and Limited Visibility Operations

A commander may desire to take advantage of the cover of darkness to gain maximum surprise or deception, maintain the momentum of successful operations, reinforce or withdraw committed units, and/or deploy maneuver support elements. The following aircraft operational requirements must be considered during night or limited visibility operations:

- Desired directions and routes of movement for aircraft (to include identification of selected terrain features).
- The identity and location of LZs and/or PZs.
- Emergency ground-to-ground signals.
- Directions and points of landing for aircraft.
- The presence of LZ obstacles is indicated to the aircraft commander through electronic/visual navigation aids.
- Illumination level.
- Moon phase/angle.
- Cultural lighting.
- Effects of shadows.

Some advantages of limited visibility operations areas are as follows:

- Aircraft are partially concealed from enemy visual observation.
- Maximum surprise and confusion can be achieved.
- Continuous pressure can be exerted on the enemy.
- Effective enemy air defense fire and interdiction by enemy aircraft are diminished.

However, disadvantages of limited visibility operations also exist. The need for more elaborate control measures and the caution required by both aviators and troops slow operations. However, with proper equipment, constant training, and a thorough knowledge of techniques, these disadvantages may be overcome. The following factors are considered:

- More time is required for planning, preparing, and executing.
- Formation flight is more difficult, and formations are more dispersed.
- LZs/PZs should be larger.
- Navigation is more difficult.
- Additional illumination is planned and must be immediately available in case it is necessary for mission accomplishment.

Operations in a Nuclear, Biological, and Chemical Environment

In the event of a nuclear attack, helicopterborne forces can conduct a radiological survey and, when feasible, move into the target area after the explosion to stall enemy exploitation of its effect. helicopterborne forces can rapidly and safely bypass obstacles created by a nuclear strike, whether their objective is within or beyond the target area.

When planning helicopterborne operations in conjunction with friendly nuclear munitions employment, the planner must consider the—

- Effects of intense light on pilot vision.
- Effects of intense heat on equipment and personnel.
- Effects of blast waves on aircraft in flight.
- Residual radiation rates on the zones.
- Utilization of LZs/PZs; however, debris may prohibit their use.
- Effects of electromagnetic pulse on electronic equipment.
- Selection of approach and retirement lanes into possible contamination LZs.
- Use of alternate LZs if primary zones have too high a residual radiation rate.

Planning for helicopterborne operations in a toxic environment also includes reconnaissance of areas known or suspected of contamination, election of routes and positions with regard to contaminated areas to avoid stirring up or spreading agents with rotor wash, and protection of supplies and equipment. The three principles of NBC operations (contamination avoidance, protection, and decontamination) are fundamental to survival during helicopterborne operations conducted in an NBC environment.

If helicopterborne operations must be conducted following contamination, the helicopterborne force may direct that hasty (spot) decontamination of aircraft be accomplished. Spot decontamination is an effective means of decontaminating specific areas of an aircraft. This sustains flight operation since certain functional areas are treated before they are touched. Surfaces must be washed with decontaminants to flush agents off the aircraft's exterior. Small amounts of the NBC agent (absorbed into the fuselage paint) will probably remain after decontaminating. The evaporation of these residues can create a vapor hazard; therefore, personnel in and around the aircraft continue to wear the protective mask and gloves. Decontamination reduces the hazard of agent contact and transfer. NBC personnel are trained in spot-decontamination procedures but may require equipment and support to effect all required decontamination quickly.

CHAPTER 6

COMBAT SUPPORT WITHIN THE HELICOPTERBORNE FORCE

Knowing combat support capabilities, assigning combat support units appropriate missions, and controlling combat support operations are essential to the application of superior combat power at the decisive time and place. The commander requires a complete understanding of MAGTF concepts in order to appropriately use combat support to enhance the combat power of the maneuver element. Combat support is organized under the GCE's command and control. In most cases, it is the GCE commander who assigns combat support units their specific missions. The GCE commander also task-organizes combat support units for movement and assigns support relationships for subordinate maneuver units.

Combat support units are normally in direct support of the helicopterborne force to ensure the close coordination and continuous, dedicated support required in helicopterborne operations. In some situations, combat support units are attached.

In most cases, combat support units are assigned to support the GCE. The GCE may assign combat support units that are attached to or in direct support if the unit could be more effectively controlled or employed by one particular unit rather than under the GCE commander's control. General support is used when combat support units can best support the operation under centralized control in order to quickly shift its efforts to the point needed. Regardless of the assigned support status, the GCE has the responsibility to ensure that combat support units are properly supported. Although the GCE is not required to provide support to the units in direct support, it is advantageous to ensure that combat support units are properly supported. This means providing rations, fuel, and ammunition as required. It also means expediting repair of equipment outside the capabilities of the GCE. This enables combat support units to provide continuous support to the GCE.

The commander of the combat support unit must be both a commander and a special staff officer. This means he commands his unit and provides advice and assistance to the helicopterborne GCE, the MAGTF commander, and the MAGTF staff. He serves as a special staff officer during the planning phase of an operation, providing assistance and advice in the preparation of the operation order (OPORD). He can also provide limited advice and assistance during the conduct of the operation, but his primary concern is command of his unit.

SECTION I. FIRE SUPPORT

Fire support is the collective and coordinated employment of mortars, artillery (to include Multiple Launch Rocket Systems [MLRSs] and high mobility artillery rocket systems [HIMARSs]), attack helicopters, CAS, naval gunfire, and other fires in support of the battle. The mission of the fire support system is to destroy, neutralize, or suppress surface targets in support of helicopterborne operations. It also includes SEAD, which is imperative in helicopterborne operations. The commander integrates the firepower of mortars, artillery, CAS, EW, and, when available, naval gunfire, with the maneuver of the helicopterborne force's combat power to–

- Destroy, suppress, and neutralize targets.
- Obscure the vision of enemy forces.
- Isolate enemy formations and positions.
- Slow and canalize enemy movements.
- Destroy, delay, disrupt, or limit the enemy at ranges greater than that of direct fire weapons.
- Screen with smoke or create obstacle areas with the employment of scatterable mines.
- Reduce the effects of enemy artillery by active counterfire.
- Interdict follow-on enemy echelons.
- Provide illumination.

To effectively utilize fire support assets, the helicopterborne force must understand artillery support relationships. The artillery commander commands his unit and serves as a special staff officer to the helicopterborne force commander during operational planning and preparation. If the task force is operating independently, it may be necessary to attach an artillery unit to provide adequate fire support. Attachment is a nonstandard mission and involves special considerations for the task force commander, such as the responsibility to provide security, logistical support, and lift capability to the artillery unit.

Fire Support Coordinator

While the helicopterborne force commander is responsible for the integration of all fires within the scheme of maneuver, the FSC is his principal assistant for the proper integration and application of fire support. The EFL and all FAC(A)s must work closely with the FSC during planning in order to ensure unity of effort and understanding during the operation. The commander and his FSC generate the maximum combat power available to support the operation.

Fire Support Delivery Means

The helicopterborne force is unique in its mission, organization, and support elements. Mission commanders must tailor the planned fire support for a helicopterborne operation specifically for that mission. Indirect fire assets must be maneuverable and capable of maintaining the rapid pace of the helicopterborne force. The fire support delivery means available to the helicopterborne force may include the following:

- Mortars, organic to each rifle company and infantry battalion, provide close-in fire support.
- Artillery must either be positioned well forward to provide fires from the PZ to the objective area or must be air lifted with the helicopterborne force to the objective.
- Supporting air defense units can provide air defense support if the situation demands and the commander directs.
- CAS will be available to provide support. Preplanned missions should be used to the maximum extent possible. CAS aircraft can be used to screen approach and retirement lanes. Because of their mobility and firepower, attack helicopters may be integrated into the fire support plan when other fire support means are not available. Mission priority for attack helicopters is to escort transport helicopters.
- Naval gunfire spot teams may be attached to the helicopterborne force if naval gunfire is available.

SECTION II. FIRE SUPPORT PLANNING AND COORDINATION

Fire support planning addresses how fire support is to be used to support maneuver forces. Fire support coordination entails those actions needed to implement plans and manage resources on the battlefield. Although planning and coordination are separated, they overlap and are mutually supporting in the fire support process. The planning and coordination process begins when the mission is received or assumed. If planning is successful, then implementation (coordination) of the plan gives the commander the support he needs to win. For more discussion on fire support planning and coordination, see MCWP 3-16, *Fire Support Coordination in the Ground Combat Element*.

The range of supporting artillery is an initial consideration when planning helicopterborne operations since helicopterborne operations may often be conducted beyond the range of artillery support. If a helicopterborne operation is beyond the range of artillery support, planners must consider prepositioning artillery forward prior to the assault or planning for additional fire support (e.g., naval gunfire, CAS, and attack helicopter) until direct support artillery can move into the area of operation.

The following factors should be considered during planning:

- LZ preparations may be fired in support of a helicopterborne assault.
- Preparations are directed against known, suspected, or likely enemy positions dominating the LZ or on the zone itself.
- The effect of fire on creating obstacles to friendly forces during landing or maneuver.
- Whether the effect of the preparation justifies loss to tactical surprise or affords the enemy sufficient time to change his tactical disposition.
- The use of fire support on the LZ itself to detonate mines/boobytraps or to reduce obstacles.
- Firing of dummy preparations to deceive the enemy.
- Possible conflict between gun-target lines and helicopter approach lanes.
- Availability of both fire support units and ammunition.
- Fires must be planned to support the consolidation of the LZ and subsequent operations.
- SEAD during both the approach and retirement of a helicopter is fundamental to its success and survival.

Formal and Informal Planning

A formal/informal planning approach at the helicopterborne force level is a combined process that uses the principles of both formal (downward) and informal (upward) planning. Initially, the FSC disseminates, in the OPORD, a fire support plan to support the force. The fire support coordination plan is modified as company plans are received. The rewritten fire plan is disseminated to supporting arms systems for execution.

Displacement of Fire Support

During the planning of fire support for a helicopterborne operation, the FSC must consider the possibility of displacement. If artillery can support the helicopterborne force's movement from a secure area (without displacing forward) then it does so. If such support is not feasible, the FSC determines if other fire support is sufficient to accomplish the mission. If other support is not sufficient, it may be necessary to displace the artillery into the objective area. Once the decision to displace is made, the following must be considered:

- Displacement is accomplished by echelon to prevent temporary loss of artillery support.
- Artillery requires security in the objective area.
- The CH-53 will probably be required to displace the artillery unit. Ammunition resupply will probably have to be done by air.
- Artillery must depend on helicopter assets for mobility unless prime movers can be lifted into the objective area.
- Supporting artillery must be available.

Enemy Air Defense

In helicopterborne operations, SEAD is a critical fire support task because assault support helicopters are especially vulnerable to enemy air defense; therefore, SEAD must be addressed during planning. Unless there are overriding tactical considerations, enemy air defense positions should always be avoided. If enemy air defense positions cannot be avoided, they must be aggressively suppressed. SEAD may be executed either as scheduled fires based on a specific time schedule or SEAD may be fired on call based on the movement of the helicopterborne force through predetermined approach and retirement lanes or across predetermined phase lines.

The FSC ensures that all flight routes and suspected enemy air defense artillery sites are targeted with preplanned fires. The FSC is normally located with the commander and requires a dedicated fire direction net in order to control the lifting and/or shifting of SEAD fires as directed by the commander. Fixed-wing or rotary-wing aircraft providing escort suppresses enemy air defenses encountered en route.

Fire Support Coordination

Fire support coordination ensures that targets are adequately covered by a suitable weapon or group of weapons. Some typical fire support coordination tasks include—

- Clearing requests for fire support.
- Selecting the best supporting arms to attack a target.
- Requesting additional fire support when needed.
- Responding to intelligence reports by requesting supporting arms to attack high-payoff targets and high-value targets.
- Coordinating the simultaneous use of different supporting arms, particularly aircraft and surface weapons.
- Synchronizing fires in support of maneuver elements.

The FSC ensures that the developed plan remains supportable. The FSC must immediately inform the commander or the S-3 if there is not enough fire support allocated to make the plan work or if changes are necessary to the plan. To do this, the FSC must be located forward with the command group during the conduct of the helicopterborne operation. For example, when a C2 helicopter is used, the FSC normally flies with the commander.

The FSC keeps abreast of the tactical situation and coordinates all fire support impacting the operation. The FSC ensures that fires do not jeopardize troop safety, interfere with other fire support means, or disrupt adjacent unit operations. During conduct of the operation, shifts in priorities of fire, changes to the fire plan to support a change in scheme of maneuver, and immediate CAS are all handled by the FSC who, in close conjunction with the S-3 and air officer, coordinates fire control activities of the helicopterborne force.

Fire support is controlled by maneuver units. Additionally, all air officers and FACs are trained to call for and adjust indirect fires. FAC(A)s and TAC(A)s can assist the helicopterborne force in coordinating or adjusting indirect fires since their location may allow them to see the battlefield.

SECTION III. ARTILLERY HELICOPTERBORNE OPERATIONS

The helicopterborne force fights both offensive and defensive battles. Therefore, its organizational tactics, which emphasize the use of artillery and aerial mobility and flexibility, require special planning considerations for employment.

Planning Considerations

Range of Artillery and Other Support Systems

Helicopterborne operations typically occur over extended distances. Therefore, the FSC must position fire support systems so they can range (place fire) and mass (concentrate fire) on targets within the helicopterborne area of operation. If the force must operate out of artillery range, there is a greater dependence on CAS and mortars.

Importance of the Target

Artillery is positioned to range those targets considered critical to the maneuver commander. For high value targets, the commander and the FSC may consider moving artillery by helicopter to strike deep in the enemy's rear by firing across the FLOT or displacing laterally in sector.

Airlift Assets

The air movement of artillery requires a heavy use of air assets. Commanders must consider the total cost of moving not just the howitzers, ammunition, and personnel, but also the maintenance and supplies needed to sustain the air-delivered artillery. These total costs must include appropriate survivability moves.

Risk in Crossing Lines

A major consideration in planning the movement of artillery in helicopterborne operations is the risk in crossing enemy lines/positions. The value of the target is weighed against the chances of survivability. Once the risk of moving artillery by helicopter is considered, the S-3 and FSC must evaluate the survivability of the artillery unit while on the ground and during extraction from the firing area.

Target Location

Artillery movement in helicopterborne operations requires pinpoint LZ and target locations. Accuracy of locations determines accuracy of fires and targets are often engaged with unobserved fires.

Pickup Zone and/or Landing Zone

Artillery displacements require PZs and LZs large enough to position equipment. When the artillery unit arrives at the LZ, it must be secure and capable of providing the unit with individual gun position(s).

Ammunition

The amount of available ammunition has a major impact on artillery support in helicopterborne operations. When planning indirect fire support, the FSC must consider the amount of ammunition required and the availability of transportation assets. Artillery ammunition supply operations place a significant burden on aviation assets available to the helicopterborne force.

Communications

In the employment of artillery, the ability to maintain communications is a requirement. The supporting unit must be within radio range of the supported unit to receive the call for fire (especially when positioning the M198 with its maximum range of 30 kilometers [km]). Unless unavoidable, firing batteries must be within communications range of their parent battalion.

Security

During helicopterborne operations, artillery relies on terrain positioning and infantry for security.

Capabilities

The only howitzer in the MAGTF inventory is the M198. The M198 has a nonrocket-assisted projectile range of 18,100 meters and its range extends to 30,000 meters with a rocket-assisted projectile. Marine forces can also anticipate the presence of MLRS and HIMARS firing in general support. These weapons can range to 60 km with rockets and 100 to 300 km with Army tactical missile systems. The CH-53E can lift the M198 and its prime mover, the M923, 5-ton truck.

Note

The CH-53E is currently rated to lift the medium tactical vehicle replacement 7-ton truck, which is replacing the M923, 5-ton truck. However, it almost reaches it performance limits when lifting the 7-ton truck. The service life extension program for the CH-53E should make lifting this vehicle a much more viable mission.

SECTION IV. AIR DEFENSE IN HELICOPTERBORNE OPERATIONS

Helicopterborne operations conducted in areas of the battlefield where the MAGTF does not enjoy air superiority will be difficult. Air superiority, according to Joint Publication (JP) 1-02, *Department of Defense Dictionary of Military and Associated Terms*, is "that degree of dominance in the air battle of one force over another that permits the conduct of operations by the former and its related land, sea, and air forces at a given time and place without prohibitive interference by the opposing force." Therefore, the effective use of limited MAGTF air defense assets becomes an important consideration in planning and executing helicopterborne operations. Since the number and type of air defense systems that can accompany a helicopterborne force are limited, and because helicopters are vulnerable to attacking aircraft and enemy ground-based air defense weapons, intelligence information must be reliable. The commander must consider the feasibility of using helicopters as the tactical situation changes. In addition to employing active air defenses, helicopterborne forces must maximize the use of passive air defense measures, such as flying at night, using terrain flight techniques, and using dispersed flight formations.

Capabilities

Ideally, effective offensive antiair warfare operations and the MAGTF's Integrated Air Defense System (IADS) provide air superiority throughout the MAGTF's area of responsibility. The ACE commander attempts to establish an IADS that provides an air defense umbrella over the entire battlefield. Some helicopterborne operations have LZs in areas that cannot be adequately covered by the MAGTF's existing IADS; therefore, it will be necessary to provide the helicopterborne unit with air defense assets specifically designated for the operation.

The scope of the operation and the air threat, as well as the availability of air defense assets, determines the type of air defense assets provided to the helicopterborne unit. In addition to dedicated air defense systems and air-to-air capable aircraft, commanders should consider the air defense capabilities of their unit's organic small arms and crew-served weapons. These weapons provide an excellent low altitude air defense capability. The key to the employment of these weapons against low-flying aircraft is coordinated, high-volume fire. See MCWP 3-11.2, *Marine Rifle Squad*, for more information concerning the use of organic small arms and crew-served weapons in an air defense role.

The employment principles for air defense weapons are essentially the same for helicopterborne operations as for other operations. Command and support relationships between air defense units and supported arms must be clearly defined. These relationships are modified, as necessary, as the operation progresses. MCWP 3-25.10, *Low Altitude Air Defense Handbook*, provides a complete description of command and support relationships, as well as the steps a commander should take in establishing an effective air defense plan.

Planning Considerations

As part of the helicopterborne operation planning process, the MAGTF commander and major subordinate commanders develop an air defense plan that supports the operation. The goal of the air defense plan is to provide continuous air defense coverage for the helicopterborne force from the time it leaves the PZs until it completes its mission. Commanders consider the following during development of the air defense plan:

- Regimental scheme of maneuver.
- Regimental commander's air defense priorities based on evaluation of each asset for criticality, recuperability, and vulnerability. See chapter 5 of MCWP 3-25.10 for a complete discussion of this evaluation process.
- Threat characteristics are used to determine the appropriate air defense system(s) to defend the specific asset:
  - Enemy location and strength.
  - Type of enemy aircraft and ordnance.
  - Past enemy attack characteristics.
  - Enemy doctrine.
- The tactical and technical capabilities and limitations of each MAGTF air defense systems.
- Aircraft employed in an active air defense role must be maneuverable.
- Surface-to-air missile (SAM) systems' (i.e., the Stinger) nighttime engagement capabilities are marginal without NVDs. Poor weather and limited visibility also limit a SAM's usefulness.
- Terrain limits tactical defense alert radar (TDAR) capabilities by causing radar masking. TDAR-equipped units should be located to minimize radar masking, while reducing signature problems (smoke, electronic, visual) as much as possible.
- Terrain and weather impact both the enemy and the effectiveness of the MAGTF's air defense weapons.
- Stinger team firing positions (primary and alternate) should provide ready access to the organic vehicle. Stinger teams require good mobility to ensure their survivability, especially after conducting an engagement that reveals their position.
- Widely dispersed, highly mobile air defense units under an IADS require centralized command (by the ACE commander through the Marine air command and control system), decentralized control (down to the lowest possible echelon), and reliable communications.

Using Helicopters to Support Air Defense Operations

In addition to supporting helicopterborne operations, Stinger units can use helicopters to occupy firing positions that are normally not accessible by wheeled or tracked vehicles. Using helicopters, Stinger units can easily cross terrain obstacles and rapidly bypass hostile areas. In addition to standard helicopter employment methods, the use of repelling, fast rope, and special patrol insertion and extraction techniques can greatly enhance a Stinger unit's ability to provide effective air defense for supported units. Using these techniques, Stinger teams can quickly deploy to sites on hilltops and other terrain features that lack adequate areas for helicopter LZs. These sites can give Stinger teams increased surveillance and overwatch capabilities, allowing them to detect and engage hostile aircraft at the maximum range of the Stinger system.

When positioning Stinger units via helicopters, commanders must consider their relative lack of mobility once they debark. Because Stinger launch signatures are highly visible, the enemy can easily locate the firing positions from which the missiles are launched. After firing, Stinger units located in particularly vulnerable positions must quickly displace to alternate firing positions. Without their organic vehicles, Stinger units are extremely susceptible to enemy counterattacks.

CHAPTER 7

COMBAT SERVICE SUPPORT

The HST provides CSS to helicopterborne operations. CSS for helicopterborne operations must be planned, organized, and executed to support a rapid tempo in highly mobile and widely dispersed operations.

Just as the helicopterborne unit is tailored to move by air, CSS must be tailored to sustain the helicopterborne unit by air. Therefore, planners must be prepared to adapt and to be innovative with available resources.

SECTION I. HELICOPTER SUPPORT TEAM

The helicopterborne unit is supported by organic, attached, and external CSSEs organized to push forces, supplies, material, and ammunition forward by air. The primary CSS organization within the helicopterborne unit is the HST. The HST is a task organization formed and equipped for employment in a LZ in order to facilitate the landing and movement of helicopterborne troops, equipment, and supplies to and within the LZ and PZ and to evacuate selected casualties and POWs. The HST performs tasks within a PZ or LZ similar to those performed by the shore party team/group in the beach support area. The functioning of the HST is the responsibility of the helicopterborne unit commander. An HST is expected to accomplish the following tasks:

- Prepare, maintain, and mark landing sites, remove or mark obstacles, and set up wind direction indicators.
- Establish and maintain required communications to include communications with supporting helicopters and supporting CSS units.
- Reconnoiter and select areas adjacent to landing sites for supply dumps and other CSS installations, HST command posts, casualty evacuation stations, and defensive positions.
- Provide LZ security.
- Direct and control helicopter operations within the LZ and support helicopter units landing in the zone.
- Provide sites for emergency helicopter repair units and refueling facilities.
- Unload helicopters (including external lifts).
- Load cargo nets, pallets, and casualties on board for return trips.
- Establish dumps, issue supplies to units, and maintain necessary records of supplies received, issued, and available.
- Provide personnel and vehicle ground control.
- Maintain situation map and information center.
- Provide emergency helicopter repair and refueling, if required.
- Evacuate POWs.
- Perform firefighting duties in the LZ.

Helicopter Support Team Organization

The HST is a task-organized unit composed of personnel and equipment of the helicopterborne force and the supporting aviation unit, with augmentation from other units as required. HST organization is determined by the operation. Normally, the HST is employed in each PZ and LZ to provide support to units operating in and around those zones. The HST normally consists of an advance party, headquarters, helicopter control element, and LZ platoon.

Advance Party

The advance party contains personnel from all elements of the HST: command, reconnaissance, communications, and LZ control. It consists of approximately 8 to 10 Marines with hand-carried equipment.

The officer in charge of the advance party makes contact with the senior Marine of the reconnaissance unit who provided the initial helicopter terminal guidance and receives a briefing of the LZ and adjacent areas. The officer in charge of the advance party assumes operational control over the HST reconnaissance unit and retains this control until the helicopter control element of the HST assumes responsibility for the helicopter control activities. Advance party personnel reconnoiter positions for the various landing sites and points to be located within the LZ. Communications personnel establish communications with the HST commander (or helicopterborne unit tactical-logistical group during amphibious operations) within the LZ, as well as communications with the helicopter unit and the helicopterborne force command post. LZ control personnel control the helicopters operating within the LZ. When the HST is established in the LZ, the advance party disbands and its personnel revert to their parent element within the HST.

HST Headquarters

The headquarters element may be provided from the service platoon of the helicopterborne unit when no CSS buildup is planned or from the landing support platoon if a CSS buildup is planned. Providing landing support personnel to the helicopterborne unit to form the HST headquarters when a CSS buildup is planned facilitates the transfer of control of the LZ to the CSS unit when the CSS buildup commences. The HST headquarters consists of—

- A command section provided by the appropriate platoon headquarters, augmented as required.
- A communications section provided by the communications platoon of the helicopterborne unit or the communications platoon, headquarters and service company, landing support battalion as appropriate.
- A military police section consisting of personnel from the military police company, division headquarters battalion or headquarters and service battalion, force service support group (FSSG) as appropriate.
- A security section provided by the helicopterborne unit to provide internal security.
- An evacuation section provided by the medical section of the helicopterborne unit.
- An HST liaison section normally accompanies the headquarters element of the helicopterborne unit.

Other MAGTF Support

The HST is task-organized to provide responsive support to a helicopterborne force. To complete its mission, the HST performs many diverse tasks in HST operations, which are normally performed by different organizations within the MAGTF. The MAGTF contributes to the mission of HST operations by providing personnel and equipment. The MAGTF organizations and the normal responsibilities of those organizations that support the HST are as follows:

- The helicopterborne unit provides overall command and control of the HST and integrates HST operations into the tactical order.
- The MAGTF command element provides the required direction and support to the helicopterborne unit.
- The MAGTF S-2 provides the intelligence necessary to plan the lift and to conduct a reconnaissance of the proposed LZ.
- The ACE provides the aircraft/aircrew, air control, and other support elements that are required by the mission.

- The GCE provides attachments, detachments, and fire support as necessary to support the helicopterborne unit.
- The CSSE provides attachments and detachments to the helicopterborne unit as necessary to ensure all CSS requirements beyond the organic capabilities of the helicopterborne unit are met.

Landing Support Battalion

The landing support battalion is a CSS unit with a unique role. It facilitates the distribution of critical, high-volume, consumable supplies throughout the MAGTF. These supplies are heavy and large and, therefore, present a difficult challenge even in normal operations. In helicopterborne operations, the difficulty of distributing these supplies is magnified.

Landing support units are present in all task-organized MAGTF CSS organizations. During amphibious operations, landing support personnel play a major role in the planning and execution of landing force support party operations. Landing support personnel play a key role in the FSSG, brigade service support group, and Marine expeditionary unit service support group. In helicopterborne operations, landing support personnel are a critical part of the HST and follow-on combat service support detachments (CSSDs) that support the helicopterborne force.

Equipment

The landing support battalion is the unit that contains the majority of the MAGTF's tactical materials handling equipment (MHE). By centralizing MHE, the MAGTF commander ensures that sufficient MHE can be quickly massed at the point of main effort; a capability that would not exist if MHE was permanently distributed among all MAGTF units. This centralization process also holds true for helicopter slings, cargo nets, and other specialized equipment used during helicopterborne operations. Centralization also ensures that there are dedicated personnel that maintain equipment in a ready condition. Centralization can also be used as an economy of force. For example, the landing support battalion is responsible for providing MHE and specialized equipment to helicopterborne units regardless of whether or not a CSS buildup is planned for a helicopterborne operation.

Training

When operations require the employment of helicopterborne units, landing support services, particularly HST services, are vital to the success of the operation. However, there are not sufficient landing support personnel within the MAGTF to perform all required HST tasks for operational success, especially in a large operation. Therefore, training of other MAGTF units by landing support units enhances the MAGTF's ability to exploit the inherent mobility of the MAGTF. All combat and combat support units must be trained and capable of moving their internal vehicles, equipment, and supplies without augmentation from the landing support battalion.

Employment

The criteria for employing landing support units and helicopterborne unit personnel to provide HST services in helicopterborne operations are not intended to limit landing support involvement, their intention is to facilitate the most efficient use of trained HST personnel. If it is determined that a helicopterborne unit lacks sufficient organic personnel to perform all of the required HST tasks, landing support augmentation should be requested even if a logistic buildup is not planned. Combat and combat support units must be capable of moving their vehicles, equipment, and supplies without augmentation from the landing support battalion.

SECTION II. COMBAT SERVICE SUPPORT PLANNING

Considerations Prior to Planning

Before planning CSS for helicopterborne operations, commanders and CSS personnel must understand the inherent characteristics of helicopterborne operations. Based on this understanding, they will implement CSS with the flexibility and prompt response time required to meet the needs of a helicopterborne operation. There are key point and specific points about helicopterborne operations that must be understood before any detailed discussion of CSS can commence.

Key Points

The CSS planner must understand the following key points of helicopterborne operations:

- Helicopterborne operations are inherently complex evolutions, requiring detailed integration of all MAGTF capabilities: aviation, ground, and CSS.
- To be effective, helicopterborne operations must be planned and executed rapidly to exploit transient enemy vulnerabilities as they occur. The MAGTF must be able to exploit these vulnerabilities before the enemy can take corrective action.
- MAGTF organizations must be trained, mentally prepared, and have SOPs in place in anticipation of the opportunity to exploit an enemy vulnerability. When the vulnerability occurs and the opportunity presents itself, the MAGTF must be capable of timely action.
- Accurate and timely intelligence is critical. Placing a helicopterborne force in the wrong place at the wrong time can result in loss of personnel, equipment, and opportunities.

Specific Points

The specific points that need to be understood by the CSS planner are as follows:

- To ensure unity of effort during a helicopterborne operation, all MAGTF units that will move to and remain in the objective area are initially attached or placed in direct support to the helicopterborne unit. The initiating headquarters of the OPORD specifies when or under what conditions control of units attached to the helicopterborne unit passes back to the parent organizations.
- Confusion that disrupts the rapid buildup of combat power into the objective area can prove to be fatal. One way to avoid confusion is to form an HST for all tactical helicopterborne operations to ensure a rapid, organized, and efficient buildup of balanced combat power in the objective area.
- An HST operation in support of a helicopterborne operation is the responsibility of the HUC. The HUC receives support and augmentation from the other MAGTF organizations to form his HST, but the responsibility for the execution of all HST tasks remains with the HUC.
- Personnel of the division or force reconnaissance units normally provide helicopter terminal guidance for the initial assault waves. ITG is especially critical for night helicopterborne operations. Once established in the LZ, the HST assumes responsibility for helicopter terminal guidance and the HUC or higher headquarters, as appropriate, assigns reconnaissance personnel follow-on missions.
- HST operations are normally terminated when the helicopterborne unit no longer depends on helicopter support as the primary means of CSS support or when a planned CSS buildup commences in the LZ.
- The helicopterborne unit and units that provide attachments to the helicopterborne units are responsible for preparing, rigging (attaching slings), and hooking up (to the helicopters) their organic equipment and supplies for external helicopter lift. This capability is acquired and maintained through training.

- Within the MAGTF, slings and cargo nets used for external helicopter lifts are centrally controlled and managed by the landing support unit. Landing support units provide training assistance in LZ operations to include external lifts to MAGTF units.
- The role of landing support battalion units in support of the HUC varies depending on the helicopterborne unit's mission. The landing support unit may be tasked by the HUC with complete responsibility for the organization and functioning of LZs/PZs or the assigned tasks may be limited to providing MHE and personnel and to providing and controlling slings for external lifts when required.
- Understanding the difference between a CSS buildup and a basic load is critical.

Planning Considerations

It is imperative that the helicopterborne unit and supporting CSS unit coordinate closely during the planning of helicopterborne operations from the initial stages onward. Concurrent planning ensures that all requirements and constraints of CSS are considered. It also provides the lead time necessary to organize and position the CSS resources required to support the operation. The HST is an essential link between the operational scheme and the CSS plan. Close and continuous coordination between the helicopterborne unit and the supporting CSS unit ensures adequate CSS throughout the operation. To organize CSS for helicopterborne operations, the CSS planner must consider the following:

- The helicopterborne unit's mission and the concept and duration of the operation.
- CSS buildup, if planned.
- The task organization (including densities of personnel, weapon systems by type, equipment by type, and aircraft by type).
- Enemy situation, weather, and terrain.
- Helicopter availability and distances between supporting and supported units.
- Ammunition, water, food, and aviation fuel consumption rates.

CSS planning must ensure that CSS is provided, not only for the organic and attached elements of the helicopterborne unit, but also for units providing direct and general support. The helicopterborne unit is responsible for planning CSS for its organic and attached units. The higher headquarters that initiates the helicopterborne operation is responsible for coordinating CSS planning of units that provide direct and general support to the helicopterborne unit. This planning must expressly designate who will provide combat support to all participating units throughout the helicopterborne operation. When an attachment joins the helicopterborne unit, the attachment brings the appropriate amounts of its own CSS assets from its parent unit. These attached assets are controlled by the HUC.

Basic Load and Combat Service Support Buildup

The basic load and the resupply of the basic load are not considered a CSS buildup. The unit brings the basic load with it and when the basic load is depleted, its own unit performs resupply. A CSS buildup takes place when supplies beyond the basic load are moved to the objective area.

Basic Load

A helicopterborne unit, based on its mission, moves to the objective area by helicopter with the necessary personnel, equipment, and a basic load of consumable supplies to accomplish the mission. The higher headquarters that assigns the mission to the helicopterborne unit also determines the helicopterborne unit's basic load.

The basic load for all classes of consumable supplies except ammunition (class V) is expressed in day(s) of supply (DOS). A DOS is the amount of supplies a unit requires to sustain itself in combat for one day. For example, a DOS for—

- Food: three meals, ready to eat per individual.
- Water: 3 to 4 gallons per individual per day in a temperate zone, but amounts are higher in both hot and cold climates.
- Fuel: the total fuel consumption of all equipment is specified in the table of authorized material. Other supplies: e.g., sandbags, barbed wire, repair parts, are normally specified on the unit SOP.

The basic load for ammunition has two parts: a basic allowance (BA) and a day(s) of ammunition (DOA). The BA is the quantity of ammunition (number of rounds) the Marine Corps has specified to be maintained by a unit for each weapon that unit employs in combat. A DOA is the total of the standard consumption rates for each organic and attached weapon when employed in combat. A DOA is further specified into an assault rate and a sustained rate. The assault rate, which is a higher consumption than the sustained rate, is specified for units conducting offensive operations. The sustained rate is specified when a unit is not conducting offensive operations. An example of a unit basic load would be one DOA assault rate and one DOA sustained rate, two DOS. The BA is always a requirement, so it is implied and normally not stated. In addition to the BA, the unit in the example will carry a DOA calculated at the higher assault rate to cover the initial assault and another DOA calculated at the sustained rate. The unit will also carry sufficient consumable supplies to sustain itself in combat for 2 days without resupply.

The basic load is issued to, controlled by, and carried by the helicopterborne unit to the objective area. The basic load is considered an organic supply to the helicopterborne unit.

Combat Service Support Buildup

Movement of the helicopterborne unit's basic load to the objective area and resupply of the basic load to maintain the specified supply level are not considered a CSS buildup. A CSS buildup occurs when supplies above and beyond the basic load are moved to the objective area. For example, if a supply safety level of one or two DOS/DOA is to be moved to the objective area, this constitutes a CSS buildup. When a CSS buildup in an LZ commences, the control of the LZ transitions from the helicopterborne unit HST to the designated CSS unit and the LZ is redesignated a landing zone support area.

SECTION III. EXECUTION OF COMBAT SERVICE SUPPORT

The helicopterborne force is normally configured to conduct the initial assault with 1 to 3 days of accompanying supplies (basic load) to ensure some degree of self-sustainment. When the enemy situation permits, resupply is accomplished by air on a routine basis to keep the basic load at the prescribed level.

CSS Trains

CSS trains for all helicopterborne units must be organized, located, and controlled to facilitate the consolidation, packaging, and air movement of the basic load into support packages configured to unit size. Generally, the air movement of the battalion's logistical train requires the same number of helicopters needed to move a rifle company. CSS trains that support a helicopterborne unit work in close coordination with the HST of that same helicopterborne unit, but the HST and CSS trains are usually separate organizations. Command and support relationships can be established between the two organizations, but since the HST is a temporary organization and CSS trains are a permanent organization, keeping the two separate promotes operational effectiveness. Certain functions in the LZ, such as distribution of ammunition and other supplies, are initially accomplished by the HST and will be assumed by CSS trains. Thus, supply personnel organic to the helicopterborne units that were initially assigned to the HST will transfer to the CSS trains when the HST is disbanded.

The organization of CSS trains varies and is based on the helicopterborne unit's mission. CSS trains may be centralized in one location (unit trains), or they may be echeloned in two or more locations (echeloned trains). In a helicopterborne operation, CSS trains normally transition between unit trains and echeloned trains.

Prior to the commencement and during the initial stages of a helicopterborne operation, unit CSS trains are employed in the vicinity of the PZ to prepare equipment and supplies for helicopter lift and to move items to the PZ. The HST takes over responsibility for final preparation and any further movement. The HST is also responsible for the initial distribution of supplies at LZs in the objective area.

As elements of the helicopterborne unit in the objective area move away from the LZ, elements of the CSS trains are echeloned into the objective area. This forward echelon assumes responsibility for receiving critical supplies contained in the unit's basic load from the HST and moving them to the elements of the helicopterborne unit that have moved away from the LZ. This echelon also provides maintenance contact teams and medical support in the objective area. CSS trains remain echeloned until such time that a CSS buildup commences, a CSSD from the MAGTF's CSSE assumes responsibility for operation of the LZ, or the HST is disbanded. If a CSS buildup commences, the responsibility for moving supplies to and issuing supplies from the landing zone falls on the MAGTF CSSE. The entire CSS train in support of the helicopterborne operation can be moved to the objective area where it will form a unit train. As operations continue in the objective area, the commander may elect to echelon the CSS trains if CSS must be collocated with maneuver units to provide immediate, dedicated support.

Supply

During the execution of CSS in support of a helicopterborne operation, the following supply issues should be considered:

- Frequent (as opposed to a few massive) replenishment of the ground and air elements. To meet this requirement, it is necessary to have a comprehensive logistical plan.
- Supplies going forward from logistical trains must be staged and moved using methods that reduce loading and unloading times. Palletized or external sling loads reduce ground time and aircraft vulnerability because they can be unloaded quickly.
- Available equipment and personnel capabilities and the anticipated load configurations must be considered when task-organizing the HST.
- The logistic plan must maintain a balance in the allocation of resources between the GCE and the ACE. This is particularly significant if FARPs are employed.
- The ability to resupply via surface methods whenever air movement is not essential to the achievement of the operational aim or if resupply by air is limited due to allocation, the enemy, or the weather. This is particularly significant if FARPs are employed.

Maintenance

Maintenance involves inspecting, testing, servicing, repairing, requisitioning, rebuilding, recovering, and evacuating equipment. Maintenance personnel do not normally accompany the assault echelon. During helicopterborne operations, repair above the operator level is accomplished in one of two ways:

- Contact teams organic to the helicopterborne unit or maintenance support teams from the supporting CSSD may be flown forward to effect immediate repair of critical equipment.
- Deadlined and/or damaged equipment may be evacuated by air.

Field and Personnel Support Services

Field and personnel support services, such as messing and billeting, are an important part of the overall support effort and continue during helicopterborne assault operations; however, these services are seldom a part of a helicopterborne assault operation. Rather, they are normally accomplished in a rear area outside the helicopterborne objective area.

Medical Support

The medical officer and the medical section of the helicopterborne unit provide medical support. To adequately support the mission, the medical officer and the chief assistant should be included in all operational and/or tactical briefings. Medical support is planned and addressed in the OPORD's administrative and logistic annex. Medical planning should include—

- Location of unit aid station in objective area.
- Ground and air evacuation plans/routes.
- Location of CSSE medical facilities.
- Location of designated casualty receiving ships or stations.
- Procedures to request helicopter casualty evacuation (including communications instructions).

Casualty Evacuation

The primary means for casualty evacuation is the helicopter. Helicopters leaving the LZ and returning to the rear area can be used to evacuate casualties. In-flight medical care is essential for those casualties whose condition is serious and must be addressed during planning. If sufficient helicopters are available, one or more helicopters may be designated as casualty evacuation helicopters for the more serious casualties who require in-flight medical treatment. The ACE normally provides in-flight medical treatment personnel. If required, augmentation can be requested from the combat support element. All casualties evacuated by helicopter are delivered to CSSE medical treatment facilities or designated casualty receiving ships if available. It is important to note that medical evacuation aircraft must be designated and properly marked to receive protections defined by the Geneva Conventions. The Marine Corps does not possess medical evacuation aircraft. Marine Corps aircraft perform a casualty evacuation mission using available combatant aircraft to evacuate casualties.

Helicopter Casualty Evacuation Control and Coordination Procedures

Procedures related to casualty evacuation are contained in the OPORD's air and medical services annexes. The medical services annex contains the medical criteria for requesting a helicopter evacuation. The air annex contains aviation-related requirements such as communications channels to request and the procedures used to control the helicopter once it enters the unit area of responsibility. Normally a unit establishes an SOP that contains both the medical and aviation aspects of helicopter casualty evacuation and the SOP is referenced in both annexes.

The helicopterborne unit establishes its unit and station near the LZ as soon as possible. During the initial stages of the operation, when maneuver units are in close proximity to the LZ, all casualties are moved to the unit aid station where minor wounds are treated and personnel return to duty if possible. The more seriously wounded are moved to the LZ where the HST evacuates them by the next available helicopter returning to the rear. Those casualties requiring in-flight medical attention are held at the aid station until a helicopter with medical personnel is inbound to the LZ. As maneuver units move further away from the LZ, it may become necessary to evacuate the more seriously wounded directly from the maneuver unit if it is possible to land a helicopter near that unit or to hoist the casualty into the helicopter if it cannot land.

The helicopterborne unit, when necessary, requests helicopter casualty evacuation from the DASC using the tactical air request (TAR) net. The request is normally initiated at a battalion FSCC once a request from the battalion aid station or a subordinate unit is received. A subordinate unit that is accompanied by a FAC may make a request over the TAR net directly to the DASC. The battalion FSCC, who is monitoring the TAR net, may disapprove the request by interrupting the transmission and voicing disapproval. Otherwise, silence is consent.

The DASC may divert an airborne helicopter, if available, to perform the casualty evacuation. If this is not possible, the DASC passes the request to the ACE's tactical air command center, which exercises launch authority. The helicopter, once airborne, receives instructions from the DASC concerning the casualty, location of pickup, flight routes, who to contact, what radio frequency to use, and the medical facility to which the casualty will be evacuated. DASC coordination with the GCE FSCC establishes a safe route through friendly fires for the helicopter.

The helicopter, when approaching the area where the helicopterborne unit is operating, contacts the FSCC of the requesting unit and receives final instruction. If the casualty pickup is made at a forward unit, the FSCC instructs the helicopter as to the radio frequency on which to contact that unit and informs the unit of the time of the helicopter's arrival. The forward unit contacts the helicopter by radio and provides terminal guidance instructions and information on the friendly and enemy tactical situation.

When time permits, identification of the casualty is reported to the battalion S-1 over the battalion administrative and logistical net.

SECTION IV. EXTERNAL LOAD OPERATIONS

Planning and executing external load operations that do not require a CSS buildup are the responsibility of the HUC, even when the HST is provided by the landing support battalion. Transporting supplies and ready for use equipment by helicopter external load (i.e., sling) has the advantage of rapidly moving heavy, outsized, or urgently needed items directly to the using unit.

The logistical planner can enhance the sustainment of the helicopterborne force by developing SOPs for sling load operations. Detailed information on the rigging of equipment and supplies for external lift by helicopter can be found in MCRP 4-11.3E, *Multiservice Helicopter Sling Load*, volumes I, II, and III.

External Load Considerations

External load considerations are as follows:

- If cargo is too light or bulky, it will not fly properly when suspended under the helicopter at cruise airspeeds.

- The external load must not exceed a helicopter's lift (under given atmospheric conditions) or hook capabilities. For general planning purposes, the following guidelines are provided:
  - CH-46 4,000 pounds.
  - CH-53D 10,000 pounds.
  - CH-53E 30,000 pounds.
  - UH-1 2,000 pounds.

Note

As outside air temperature and/or altitude increases the payload capacity of a helicopter decreases.

- Airspeeds may be slower when helicopters carry external loads.
- Dust, sand, or snow, which would be blown during hover for pickup or delivery of cargo, may preclude safe external load operations.
- Higher altitudes, which may be flown with sling loads, may subject the aircraft to more ground fire.
- External hovering to pick up or deliver a sling load during darkness is inherently more dangerous than similar daylight operations.
- The availability of suitable sling, cargo nets, cargo bags, and other air delivery items may preclude or limit external load operations.

Elements of an External Lift Mission

There are normally three different elements involved in an external lift mission: the PZ HST, the LZ HST, and the aviation unit.

The PZ HST is responsible for—

- Preparing and controlling the PZ.
- Repositioning all the equipment needed for external lift operations, including sling, A-22 bags, cargo nets, and containers.
- Storing, inspecting, and maintaining all external lift equipment.
- Providing a sufficient number of trained HST crews for rigging and inspecting all the loads, guiding the helicopters, hooking up the loads, and clearing the aircraft for departure.
- Securing and protecting sensitive items of supply and equipment.
- Providing load derigging and disposition instructions to the receiving unit.
- Providing disposition instructions to the receiving and aviation units for slings, A-22 bags, cargo nets, and containers.

The LZ HST is responsible for—

- Preparing and controlling the LZ.
- Providing trained HSTs to guide the aircraft in and de-rig the load.
- Coordinating with the PZ HST for the control and return of the slings, A-22 bags, or any other items that belong to the supported unit as soon as possible.
- Preparing, coordinating, and inspecting backloads (e.g., slings, A-22 bags) and having them ready for hookup or loading.

The aviation unit is responsible for—

- Effecting and/or establishing coordination with the helicopterborne unit.
- Advising the helicopterborne unit on the limitations of the size and weight of the loads that may be rigged.
- Advising the helicopterborne unit on the suitability of the selected PZs/LZs.
- Providing assistance for the recovery and return to the PZ of the slings, A-22 bags, cargo nets, and containers as required by the supported unit.
- Establishing safety procedures that ensure uniformity and understanding of duties and responsibilities between the ground crew and the flight crew.

SECTION V. AVIATION SUPPORT CONSIDERATIONS

Aviation units consume large amounts of fuel, ammunition, class IX supplies, and maintenance support during intensive helicopterborne operations. Although aviation units are responsible for meeting their own unique logistical support requirements, the MAGTF planner must be aware of the requirements, plan for them, and be prepared to assist as necessary.

Forward Arming and Refueling Points

FARPs are temporary facilities. They are organized, equipped, and deployed by the ACE commander. FARPs are positioned in or closer to the area of operations than the aviation unit's combat service area. The FARP permits combat aircraft to rapidly refuel and rearm simultaneously. FARPs are—

- Established in the vicinity of the supported ground unit. Whenever possible, these will be established behind the FEBA/FLOT, and out of range of the majority of enemy artillery units.
- Hasty and mobile FARPs are often established forward of the FEBA/FLOT. Because of their short duration and mobile nature, they are less likely to be targets of enemy artillery and attack.
- Positioned to reduce turnaround time, thus optimizing helicopter availability.
- Repositioned frequently to avoid detection and destruction.
- Fully mobile, using ground vehicles and helicopters.
- Capable of performing refueling and re-arming operations rapidly and efficiently.
- Defended from enemy ground and air attack.
- Concealed from observation.

Aircraft Maintenance

Aircraft have substantial maintenance requirements. However, maintenance is kept to a minimum in the operational area. A method used to accomplish this and still have responsive maintenance is to move aviation maintenance teams to the aircraft requiring repair when the repair is beyond the capability of the aircraft crew. The ACE commander may assign aircraft maintenance teams to accompany the flight or position them in PZs and LZs.

Tactical Recovery of Aircraft and Personnel

If an aircraft is forced to land on enemy terrain due to mechanical problems or combat damage, every effort is made to protect the aircraft and personnel until they can be evacuated. However, mission execution has priority over rescue and recovery operations. The ACE commander is notified immediately of any downed aircraft. He takes action in accordance with unit SOPs to secure and recover personnel and aircraft with resources or requests assistance from the MAGTF commander. The helicopterborne unit or other MAGTF unit may have to provide security for recovery teams.

When an aircraft is downed, the senior occupant assumes command and establishes a defense of the area or organizes evasive action. If an aircraft is abandoned, steps are taken to destroy it to preclude its capture or the capture of sensitive equipment or documents. The level of authority required to destroy the aircraft is established in unit SOPs (it may be covered in the OPORD). However, if capture is imminent, the aircraft, equipment, or documents should be destroyed.

CHAPTER 8

CONDUCT OF A HELICOPTERBORNE OPERATION

This chapter addresses the HTF's movement from the assembly area to the PZ and on to the LZ. The helicopterborne operation, during subsequent operations ashore, normally begins at an assembly area. If subsequent lifts are required in the same operation, the procedures described in this chapter are repeated. Planning starts from the assembly area and progresses through the final objective. If any extraction is required, LZ(s) in the vicinity of the objective area are determined during the initial planning phase.

Note

This discussion is not all-inclusive, certain actions may be omitted or added as required by operational demands.

Movement From the Assembly Area to the Landing Zone

At the prescribed time, units move from the assembly area to the holding area via a route designated by the HUC. A holding area must be—

- Covered and concealed.
- Sufficient size for the helicopterborne force.
- Close to primary and alternate PZs.

Each unit commander notifies the PZCO upon his unit's arrival in the holding area. In the holding area, unit leaders separate the unit into loads (sticks) according to the loading plan. Heavy loads and external loads should not be programmed in initial waves. Offloading heavy internal loads is time-consuming and slows troop buildup.

Each load (stick) includes a designated helicopter team leader. The helicopter team leader is usually the senior Marine on the helicopter team and is responsible for briefing his troops and inspecting the load. The helicopter team leader ensures that the load is organized and ready to be loaded as planned. Upon arrival at the holding area, the PZ control party briefing includes the loading point for primary and alternate PZs and the routes to those points. At a minimum, the helicopter team leader briefs his helicopter team members on the following information:

- Loading procedures.
- Bump plan (individual/load bumps).
- Safety belt usage.
- Preflight safety inspection of Marines.
- In-flight procedures.
- Downed aircraft procedures.
- Offloading procedures.
- Movement from the LZ.

Procedures in the Pickup Zone

Organization of the Pickup Zone

To the maximum extent possible, the PZCO lays out the PZ as directed in the plan.

To minimize confusion during landing, aviation elements arrive at the PZ in the formation directed in the plan. Then, the PZCO, or HST personnel, assists in loading by ensuring helicopters and personnel are in the proper location and formation at the correct time. If an aircraft (scheduled for the lift) is unable to complete its mission due to mechanical failure, the PZCO informs the unit commander, who implements the bump plan.

Infantry Movement to the Pickup Zone

The PZCO coordinates the arrival of both aircraft and troops so that the troops arrive at their respective loading point just before the helicopters land. This prevents congestion, facilitates security, and reduces vulnerability to enemy actions during PZ operations.

On the PZCO's signal, loads (sticks) move by designated routes from their holding area to their loading points in the PZ. The PZCO may use schedules, messengers, hand-and-arm signals, light signals, or (as a last resort) radio to order loads to move to the PZ.

Helicopter Movement to the Pickup Zone

Helicopters begin movement to arrive in the PZ at the scheduled time and should use terrain flying techniques en route to the PZ. The PZCO contacts the aviation unit if there is a PZ change. If there has been a change in allowable lift/load, number of aircraft, or the landing formation, the AFL must contact the PZCO.

During air movement to the PZ, enemy antiaircraft or other fire may be encountered. Therefore, air reconnaissance may be used to locate and suppress enemy positions prior to the arrival of the helicopterborne assault aircraft. Attack helicopters will not normally land in the PZ. When assault helicopters are to be on the ground for extended periods, attack helicopters may occupy holding areas nearby or return to FARP sites. The C2 helicopter is positioned where the command group can see and control critical events.

Strict radio discipline is maintained throughout the operation; radio silence should not be broken unless absolutely necessary. Radio calls between aircraft are permitted only as a last resort when other signals are not appropriate. Use of covered frequency-agile nets further reduces the requirement for radio silence.

Lift-off From the Pickup Zone

When the assault aircraft are loaded and ready for lift-off, the PZCO signals the flight leader using hand-and-arm or light signals. The AFL may signal other aircraft by turning his navigation lights on (or off). Members of the PZ control group may also relay the alert to lift-off to aircraft in the rear of the formation or the flight leader may lift-off and the others follow.

Lift-off should be at the time prescribed in the plan. However, aircraft will not loiter in the PZ. If they are ready early, they lift-off and alter speed so as to arrive at designated locations at the appropriate times. This should place the first aircraft of the first lift in the LZ at L-hour.

Lift-off may be by single aircraft or by wave. Under some conditions (e.g., dusty PZ, restricted PZ, or high density altitude and no wind), it is best to break waves into smaller increments. If LZ insert is executed in a single wave, then simultaneous lift-off is preferred because—

- It is easier for the EFL to plan the scheme of maneuver and provide security en route for aircraft depending on number of escort aircraft available.
- Operational control is easier.
- It reduces the enemy's time to fire at the aircraft.
- The AFL adjusts the flight's speed and rate of climb so all elements form into the en route flight formation at the required altitude.

En route to the Landing Zone

The AMC predetermines the en route flight speed. The AFL paces the flight to ensure the flight crosses the line of departure on time.

Communications security is paramount unless using covered nets. However, if directed in the order, flight leaders report to the AMC and mission commander as they pass each checkpoint, especially when checkpoints are tied to triggers such as fires. Checkpoint information must be passed to the helicopter team leader.

Ground commanders, helicopter team leaders, and aircraft crewmembers must remain oriented throughout the flight. To remain oriented, they use the aircraft's internal communications system, which receives information updates from the aircrew, and they follow and verify the flight route using terrain observation, maps, aircraft compass, and aircraft speed.

When a threat is encountered along the flight route, which prohibits the helicopterborne force from using that route, the AFL gives the order for the AMC or mission commander to modify or switch to alternate flight routes. This authority may be delegated to the AFL. If the LZ needs to be changed, the HUC makes the decision and informs the AMC or mission commander. The AMC and AFL may also be given authority to change the LZ based on the enemy threat or hazardous environmental conditions.

Security

Attack helicopters provide security for downed aircraft, route reconnaissance, and other assistance en route as directed by the ACE commander. The ACE commander develops the plan for TRAP.

Fixed-wing aircraft may work with attack helicopters to provide security to the flank, front, and rear of the helicopter formation(s). When performing this role in a medium to high threat environment, specially-equipped aircraft suppress or destroy SAM sites and radar-directed guns. Other fixed-wing aircraft may be used to selectively jam enemy radar and communications signals using jamming transmitters or other methods.

When available, indirect fire weapons provide suppressive fires along the flight routes as planned or as necessary.

Landing Operations

Attack helicopters can be employed in various roles during landing operations. They may—

- Precede the assault element into the LZ for reconnaissance and/or provide suppressive fires to prevent a time gap in LZ fires (provided by other support elements). The EFL also determines if the criteria that will permit successful insertion of the GCE exists. This assessment is based on destruction criteria of threat forces established during planning.
- Recommend last minute changes regarding aircraft landing instructions.
- Provide area cover, neutralize known enemy positions, or provide security for assault aircraft while in the LZ area.
- Observe ground approaches to the LZ for possible enemy attacks.

Command and Control Helicopter

The C2 helicopter is positioned where it can best observe and communicate with the forward elements. The HUC determines where he can best influence the action by either remaining in the C2 aircraft or by joining forces on the ground.

Landing Zone Preparatory Fires

Preparatory fires are planned for all primary and alternate LZs. The decision to initiate LZ preparatory fires is made by the mission commander. The mission commander can delegated this authority to the AMC, EFL, HUC, FSC, or operations officer. The FSC should travel with the ground commander to expedite fires and changes to preplanned fires. To the maximum extent possible, fires are planned along all routes leading to the LZ. Planned fires should be intense and should shift or lift shortly before the first elements land. In the development and sequencing of fires, the following are considered:

- Deception fires, while not fired on the objective area, should still be fired against targets of some tactical value; however, economy of force must be considered.
- Preparations of a long duration may reduce the possibility of surprise.
- The FSC considers the availability of fire support assets and coordinates their use with artillery units. Preparations by fixed-wing aircraft are requested through the FAC(A).
- A known or suspected enemy force located in the vicinity of the LZ, regardless of size, warrants LZ preparation if the LZ is to be used. The GO/NO GO, LZ cold, and destruction criteria should be established based on the threat.
- Various types of ordnance used in preparation fires can cause obstacles to landing and maneuver (e.g., craters, tree blow-down, fires, smoke, poor visibility) on and near the LZ.
- Fire support coordination measures must be established for lifting or shifting fires; e.g., restrictive fire line or restrictive fire area.
- Resupply ammunition limitations.

Because CAS on-station time is limited by fuel and enemy air defenses, the sequencing of supporting fires is carefully controlled by the FSC to obtain maximum, continuous support. To ensure that all fire support assets are utilized at the correct time, the FSC is collocated with the DASC and must be constantly informed as to the status of the flight. This allows fires to coincide with the actual arrival of landing helicopters at the LZ.

Another method of continuing fire support is to shift indirect fires to one flank, conduct a simultaneous air strike on another flank, and use attack helicopters to orient on the approach and retirement lanes. This technique requires precise timing and helicopter formation navigation to avoid flight paths of other aircraft and gun-target lines of indirect fire weapons.

Landing Techniques

The helicopterborne force should land as planned unless last minute changes in the tactical situation force the commander to abort or alter the landing. The aircrew must make every effort to keep the troops in their aircraft informed of the situation, especially of any changes to the original plan.

A simultaneous or near simultaneous landing is desired so as to place the maximum number of troops on the ground, in a given area, in the shortest possible time. Individual aircraft landing points are planned to disembark troops as close as possible to their initial positions. If ground movement times are a factor, staggered waves or landing by element are used to reduce LZ size and the time required for ground movement.

In most operations, if the situation permits, the operation is accomplished with a minimum number of lifts, each with the maximum number of aircraft that the LZ will accommodate. This reduces the exposure time of the aircraft, maintains unit integrity, provides maximum combat power, and gives the enemy less time to react. When separate element landings are dictated because of LZ size, time intervals between elements are kept as short as possible. Detailed planning determines the minimum time needed between waves of assault aircraft, facilitates the safe insertion of forces, and facilitates the ground scheme of maneuver.

Troops are most vulnerable during landing; they disembark rapidly and deploy to carry out assigned missions. Therefore, planning must maximize suppressive fires provided by assault door guns during disembarkation.

Casualty evacuation locations are normally designated at the approach end of the LZ. This permits continuation of the lift and prompt evacuation of the wounded.

At the LZ, leaders at all levels account for personnel and equipment and submit appropriate reports to higher headquarters. Key personnel killed, wounded, or missing are replaced according to unit SOP. Essential weapons missing or out of action may require the force to reorganize. After the unit completes its consolidation of the LZ, it reorganizes as necessary to carry out the ground tactical plan.

Completion of the Landing Zone Operation

When the LZ operation is finished, aviation assets return by preselected routes to complete subsequent lifts, conduct other operations, or refuel and remain in support of ground forces (e.g., casualty evacuation, immediate re-embarkation, emergency extract, resupply, TRAP).

Commanders' Responsibilities/ Sequence of Actions

Helicopterborne Mission Commander or Helicopterborne Unit Commander

The helicopterborne mission commander or HUC takes the following actions:

- Receives warning order (see app. I).
- Conducts mission analysis.
- Receives initial information from the AMC.
- Gives warning order to staff and subordinates.
- Receives personnel status report from S-1.
- Receives equipment status report from S-4.
- Receives enemy situation briefing from S-2.
- Receives friendly forces information briefing from S-3.
- Continues analysis of METT-T.
- Receives higher headquarters' OPORD.
- Begins development of commander's estimate.
- Provides guidance to staff as needed.
- Receives staff estimates.
- Obtains data from staff as needed.
- Announces concept.
- Supervises development of OPORD.
- Receives air movement information.
- Coordinates air movement matters with AMC.
- Receives air loading plan from S-3.
- Receives copy of OPORD from S-3.
- Approves or modifies and approves OPORD.
- Issues or oversees issuance of OPORD.
- Conducts/oversees conducting of OPORD brief.

Air Mission Commander

The AMC takes the following actions:

- Receives warning order.
- Conducts mission analysis:
  - Receives aircrew status report from ACE S-3.
  - Receives aircraft availability report from ACE aircraft maintenance officer.
  - Receives enemy situational brief from ACE S-2.
- Gives initial planning information to GCE and staff.
- Receives GCE warning order.
- Receives friendly force information briefing from S-3.
- Provides technical advice to GCE executive officer (XO) and S-2 for PZ/LZ identification.
- Coordinates with supported unit staff.
- Provides information to aviation unit on ground operation.
- Provides advice to GCE S-3 on PZ selection.
- Assists XO in PZ control plan.
- Provides GCE air officer with flight route computations.
- Provides advice to GCE S-3 on LZ and flight route selection.
- Coordinates PZs/LZs, flight routes, and aircraft allocation from GCE air officer and HUC.
- Obtains PZ control plan from GCE XO (PZCO).
- Aids GCE S-4 in selecting logistic PZ(s).
- Coordinates aircraft internal and sling equipment loads with GCE S-4.

- Obtains HWSAT from GCE S-3 and develops the HEALT in accordance with the S-3 and HUC.
- Briefs aviation unit on operation.
- Inspects PZ(s) with GCE XO.
- Receives GCE OPORD.

Helicopterborne Force Executive Officer

The following actions are taken by the helicopterborne force XO:

- Receives warning order.
- Receives personnel status report from S-1.
- Receives equipment status report from S-3.
- Receives enemy situation report from S-2.
- Receives AMC initial information.
- Receives friendly forces information from S-3.
- Determines available PZs. Obtains advice from AMC.
- Submits PZs to S-3.
- Coordinates staff planning.
- Obtains PZ from S-3.
- Develops PZ control plan.
- Coordinates PZ operations with AMC/terminal controller(s).
- Receives GCE commander's concept.
- Obtains PZs, LZs, flight routes, and aircraft allocation from S-3.
- Coordinates PZ operations with S-1. Completes bump and straggler control plan.
- Inspects PZs with PZCO(s).
- Obtains air movement plan from S-3.
- Obtains air loading plan from S-3.
- Obtains sequence of bump from subordinate units. Annotates air movement plan with sequence of bump.
- Completes PZ control plan. Submits to S-3.
- Inspects PZs with mission commander.
- Receives OPORD.

Helicopterborne Force S-1

The following actions are performed by the helicopterborne force S-1:

- Receives operation notification.
- Assembles personnel data.
- Receives helicopterborne warning order.
- Reports personnel status to commander and staff.
- Receives mission commander's initial information.
- Receives friendly forces information briefing from S-3.
- Begins mission analysis from personnel standpoint.
- Begins preparation of staff estimate (personnel).
- Receives commander's concept.
- Coordinates PZ operations with battalion XO. Develops straggler control plan.
- Briefs subordinate unit personnel on straggler control plan.
- Receives command post's general location from S-3.
- Coordinates POW and civilian control plan with S-2.
- Completes POW and civilian control plan. Coordinates with S-4.
- Completes S-1 portion of paragraph 4, OPORD. Gives to S-4.
- Receives air loading plan from S-3.
- Coordinates with headquarters commandant. Develops plan for command post displacement and security.
- Coordinates command post displacement plan with S-3.
- Coordinates with S-2 for interpreter support, if applicable.

- Coordinates casualty evacuation plan with S-4, surgeon, and AFL.
- Plans for recovery and evacuation of the dead. Coordinates with S-3 and S-4.
- Develops personnel replacement plan, if applicable.
- Completes Annex E (Personnel) of the OPORD.
- Receives OPORD.

Helicopterborne Force S-2

The helicopterborne force S-2 takes the following actions:

- Receives operation notification.
- Receives command direction regarding mission, area of operations.
- Coordinates any needed map requests through S-4.
- Assembles available intelligence data.
- Requests weather forecast.
- Distributes maps.
- Provides initial intelligence orientation and brief to commander, staff, and subordinate unit commanders on enemy situations.
- Obtains advice concerning LZs from AMC.
- Determines available LZs.
- Submits LZ list to S-3.
- Analyzes weather forecasts.
- Obtains advice concerning flight routes from air officer and AMC.
- Develops threat data information for proposed flight routes and provides data to AMC and HUC.
- Provides threat estimate along with available flight routes to S-3 (air officer).
- Develops escape and recovery plans; coordinates with ACE S-2.
- Recommends priority intelligence requirements and information requirements.
- Develops intelligence collection plan with ACE and briefs S-3 on plan.
- Tasks available collection assets and recommends employment of reconnaissance assets to the S-3.
- Submits requests for intelligence to higher authority.
- Requests aviation reconnaissance and/or imagery of routes, LZ(s), and objectives.
- Begins preparation of staff intelligence estimate.
- Processes intelligence data gathered.
- Prepares intelligence products.
- Develops intelligence debrief plan.
- Completes staff estimate (intelligence).
- Provides S-3 with staff estimate (intelligence).
- Continues intelligence cycle and provides regular briefings to commanders and staffs.
- Recommends targets to FSC and S-3.
- Completes paragraph 1b2 (Area of Interest), 1c (Enemy Forces), and Annex B (Intelligence) of OPORD and submits to S-3.
- Continues intelligence cycle and provides regular briefings to commanders and staffs.
- Coordinates with S-1 for interpreter support, if applicable.
- Coordinates development of POW and civilian control plan with S-1.
- Develops debrief plan.
- Updates intelligence map as needed.

Helicopterborne Force GCE S-3

The helicopterborne force GCE S-3 takes the following actions:

- Receives warning order.
- Assembles data on friendly elements.
- Receives mission commander initial information through the air officer.
- Receives personnel status from S-1.
- Receives equipment status from S-4.
- Receives enemy situation from S-2.
- Briefs mission commander's initial information.
- Briefs friendly forces disposition and location.
- Begins development of course of action.
- Receives brief by S-2 on collection plan.
- Receives list of available LZs from S-2 and available PZs from XO.
- Receives higher headquarters' OPORD.

- Begins preparation of staff estimates for operations.
- Selects PZs. Briefs XO and air officer on PZ selection.
- Obtains available flight routes from mission commander through air officer.
- Consolidates staff information.
- Recommends LZs and flight routes.
- Determines need for indirect fire preparations.
- Determines need for EW support.
- Provides staff estimates of supportability to commander.
- Receives commander's decision.
- Begins preparation of OPORD.
- Provides XO and air officer with PZs and aircraft allocation. Selects general location for command post. Provides information to staff.
- Receives S-2 input to OPORD. Receives administrative-logistical portion of order from the S-4.
- Completes OPORD paragraphs 1, 2, 3, and 5 and Annex C (Operations). Supervises completion of pertinent annexes and assembles completed OPORD.
- Receives air movement plan from air officer.
- Receives fire plan from FSC.
- Receives EW support plan from communications officer.
- Obtains air loading plan from air officer.
- Receives paragraph 4 of the OPORD from S-4.
- Completes paragraph 5 of the OPORD.
- Completes operation overlay.
- Coordinates command post displacement with headquarters commandant.
- Obtains PZ control plan from XO.
- Complete OPORD with annexes. Submits to commander for approval.
- Receives OPORD from the commander.
- Issues OPORD, when directed by the commander.

Helicopterborne Force GCE Air Officer

The helicopterborne force GCE air officer takes the following actions:

- Receives operation notification.
- Receives helicopterborne warning order.
- Receives personnel status from S-1.
- Receives equipment status from S-4.
- Receives enemy situation briefing from S-2.
- Receives mission commander's initial information.
- Analyzes mission commander's initial information for available assets.
- Establishes liaison with ACE/AMC.
- Assists S-3 in preparation of air movement plan.
- Recommends air requests to S-3 and processes air requests from S-3.
- Obtains PZs from S-3 and provides PZs to FSC and staff as needed.
- Establishes necessary liaison with tactical air control party and coordinates preplanned fire support.
- Receives available flight routes from S-2.
- Computes flight route times and distances.
- Provides available flight route information to the S-3.
- Receives helicopterborne commander's concept.
- Obtains LZs, flight routes, and aircraft allocation from S-3. Provides data to mission commander, FSC, subordinate unit commanders, and staff as needed.
- Obtain any additional tactical air requirements from FSC.
- Initiates requests for air support of all types (e.g., assault support request, joint TARs).
- Obtains logistical PZs from S-4.
- Coordinates air movement plan with HUC and AMC and submits to S-3 for approval.
- Distributes air movement plan.

- Obtains subordinate air loading plan.
- Consolidates air loading plans and provides to helicopterborne commander, S-3, XO, and S-1.
- Receives OPORD.

Helicopterborne Force GCE S-4

The helicopterborne force GCE S-4 takes the following actions:

- Receives operation notification.
- Obtains maps requests by S-2.
- Assembles equipment data.
- Receives helicopterborne warning order.
- Receives personnel status from S-1.
- Provides equipment status to the commander and staff.
- Receives enemy situation briefing from S-2.
- Receives mission commander's initial information.
- Receives friendly forces information from S-3.
- Conducts mission analysis to determine logistic/sustainment requirements.
- Receives initial supply requirements from subordinate units.
- Begins preparation of staff estimate (logistics).
- Determines effects of ammunition supply rate on operations. Submits ammunition supply rate to FSC.
- Compiles material usage data for operation. Obtains PZs from S-3 (air officer).
- Compares usage data to materiel available.
- Requests materiel as needed.
- Coordinates with mission commander and AMC on establishing FARP.
- Provides S-3 with staff appraisal (logistics).
- Receives helicopterborne commander's concept.
- Begins development of support plan for operation. Obtains LZs and flight routes from S-3.
- Selects logistic PZs and provides to S-3 (air officer).
- Plans aircraft loads (internal and external) for mission support. Coordinates pickup points with mission commander and/or S-3 air officer.
- Completes paragraph 4 and Annex D (Logistics/Combat Service Support) of the OPORD.
- Coordinates plans for evacuation of enemy materiel with S-2.
- Coordinates casualty evacuation plan with S-1, medical, and AMC.
- Receives OPORD.

Helicopterborne Force GCE Fire Support Coordinator

The helicopterborne force GCE FSC takes the following actions:

- Receives operational notification.
- Begins mission analysis to determine available and needed means of fire support.
- Plots locations and capabilities for indirect fire support systems supporting the force.
- Estimates fire support needed.
- Obtains ammunition supply rate from S-4/S-3 of artillery unit, determines effects of ammunition supply rate on operation, and gathers information for development of fire support plan.
- Obtains PZs from S-3.
- Continues to gather information for development of fire support plan.
- Coordinates fire support requirements with S-3.
- Provides S-3 with available fire support recommendation for indirect fire preparations.
- Obtains LZs and flight routes from S-3.
- Obtains recommended targets from S-2. Develops air requests to support ground tactical plan.

- Completes fire support plan (completes paragraph 3f[3] [Fire Support Coordinating Measures], paragraph 3t [Fire Support], and Appendix 19 [Fire Support] of Annex C of the OPORD).
- Submits fire support plan to S-3 for commander's approval; on approval, distributes.

Note

Subordinate units develop their fire support plans. The FSC coordinates and consolidates subordinate unit plans into the helicopterborne plan.

- Receives OPORD.

Subordinate Unit Commander

The subordinate unit commander takes the following actions:

- Receives operational notification.
- Gathers personnel and equipment data.
- Reports personnel and equipment to battalion staff. Receives maps.
- Receives battalion warning order.
- Issues company warning order.
- Determines initial supply requirements.
- Submits initial supply requirements to S-4.
- Begins preparation of air loading plans.
- Continues mission preparation.
- Obtains appropriate PZs, LZs, flight routes, and aircraft allocation from S-3.
- Continues mission planning.
- Obtains air movement plan from S-3.
- Completes air loading plan.
- Submits air loading plan to S-3.
- Receives OPORD.
- Analyzes mission.
- Develops fire support plan.
- Develops ground tactical plan.
- Prepares OPORD.

Helicopterborne Force Commander

Planning Sequence

The commander carefully analyzes the tasks and elements that are required to accomplish the mission and maintain unit integrity. The commander must also consider the five basic plans that comprise the helicopterborne force operation. These plans—the ground tactical plan, the landing plan, the air movement plan, the loading plan, and the staging plan—are developed concurrently. The ground tactical plan is driven by the assigned mission and is, therefore, developed first. Consequently, it forms the basis from which other plans are derived.

Time Schedule

Planning for the helicopterborne force operation requires time—time to plan, time to prepare, and time to brief. The planning is as detailed as time will permit.

The HUC, Commanding Officer, 3d Battalion, 6th Marines, receives the Marine expeditionary brigade (MEB) warning order at 0900. He determined that his force would be ready to land, L-hour, at 0600 the next morning. At 0945, he met with his staff and officers at the battalion command post and issued his warning order. Using the reverse planning sequence, the HUC outlined the following sequence:

0600	L-hour
0530	1st assault wave departs PZ
0515	En route from assembly area to PZ
0500	Units arrive assembly/staging area
0415	Reveille
2100	Battalion XO's brief
2000	Status update from battalion staff
1800	Evening meal
1700	Issue OPORD X-9X (JUSTSAYNO)
1600	Receives S-2 intelligence brief
1500	Receives S-4 equipment and logistical brief
1400	Receives S-1 personnel status brief
1300	Receives S-3 brief–Consolidate–S-2 brief goes first
1200	Noon meal
1100	Complete issuance of warning order
0945	Issue warning order

Ground Tactical Plan

All planning evolves around the ground tactical plan. The plan specifies actions in the objective area that ultimately accomplish the mission.

> The battalion commander is faced with three primary objectives—
>
> - Primary objective number 1: LZ SNOWBIRD.
> - Primary objective number 2: Objective Z, the Ande Municipal Airport.
> - Primary objective number 3: Link up with the mechanized force.
>
> The battalion commander is determined to keep the operation as simple as possible. Therefore, he assigned one mission to each of his rifle companies.
>
> - Company I (reinforced) would secure Primary Objective Number 1, LZ SNOWBIRD. Company I would provide security at the LZ and guidance to all incoming assets. The company commander would exercise control over the LZ, provide guides for the incoming units, and maintain security to preclude paramilitary forces from disrupting the landing plan. Once Company L has arrived, Company I would become the battalion reserve.
> - Company K (reinforced) would follow Company I into LZ SNOWBIRD and immediately deploy to seize Objective Z, the Ande Municipal Airport. Company K would continue operations until it has secured complete control of the air facility. This control would be established to allow 6th Marine Expeditionary Force (MEF) and the government of Grande to operate from the airstrip and use the buildings.
> - Company L (reinforced) would initially act as the reserve unit. Company L would help the designated PZCO, Commanding Officer, Headquarters and Service Company. Company L would provide security for the PZ and personnel as needed to assist units moving from the staging/assembly area to the PZ, as well as help load materiel into aircraft as needed. On order, Company L would load at the PZ, land in LZ SNOWBIRD, and conduct the linkup operation.
>
> A warning order was given to the battalion staff and company commanders at the 0945 meeting.

Landing Plan

The landing plan must support the ground tactical plan.

> The commander examined the following:
>
> - Helicopter assets were available to enable two reinforced companies to be airlifted simultaneously. The battalion commander decided to lift Company I (reinforced) en mass. They would land in LZ SNOWBIRD at 0600 and secure it. Fifteen minutes later, Company K (reinforced) would land in a single wave.
> - Once Company K had departed the LZ, the rest of the battalion could commence air movement. Initially, the battalion command post would land with Company K and set up in the vicinity of LZ SNOWBIRD. On signal, the battalion command post would establish itself in the vicinity of the Ande Municipal Airport.
> - At the conclusion of landing operations, Company I would remain in the areas adjacent LZ SNOWBIRD.

The landing plan sequences elements into the area of operation to ensure that units arrive at the designated location at the designated time to execute the ground tactical plan.

> The following considerations were examined and decisions were made.
>
> - Size and location of LZ.
> - Anticipated forces in and around the LZ.
> - Unit tactical integrity.
> - Ensuring all Marines are briefed and oriented.
> - Ensuring Company I is sufficiently task-organized and equipped to destroy the enemy in the area and secure the LZ.
> - Ensuring the landing plan offers flexibility in the event circumstances require it.
> - Planning supporting fires in and around the LZ.
>   - Plan fires for air movement.
>   - Plan fires for the landing.
>   - Plan fires for subsequent operations.
> - Ensuring plans were made for resupply and casualty evacuation.

Air Movement Plan

The air movement plan is based on the ground tactical plan and the landing plan. It specifies the schedule and provides instructions for air movement of troops, equipment, and supplies. This plan is detailed in the HEALT. Furthermore, it provides coordinating instructions regarding air routes, control points, speeds, altitudes, and formations. The planned use of aviation fire support, security, and linkup operations should be included.

> **Air movement for this operation were developed by the air officer in coordination with the ACE.**
>
> **Tentative flight routes were selected by the AFL.**

The air movement plan is prepared jointly by the GCE and the ACE. The air movement plan contains aircraft allocations; designates the number and type of aircraft for each wave of the operation; specifies departure points; identifies routes to and from the PZ and LZ; and identifies the loading, liftoff, and landing times. The air movement plan ensures that all required personnel and materiel are accounted for in the movement and that each aircraft is properly loaded, correctly positioned, and directed to the LZ.

Loading Plan

The loading plan is based on the air movement plan and detailed in the HEALT and HWSAT. Unit integrity is essential, however, personnel weapons and equipment may be spread loaded so that C2 assets, combat power, and an appropriate weapons mix arrive in the LZ ready for combat. A bump plan ensures that essential personnel and equipment are loaded ahead of less critical loads in case of aircraft breakdowns or delays.

> **The loading plan for 3d Battalion, 6th Marines was contained in the battalion SOP for helicopterborne operations. Load plans were carefully coordinated with the aviation elements and verified by the embarkation officer and the air officer. The loading plan would control the movement of troops, supplies, and equipment at the PZ; designate unit loading sites; and control the arrival, loading, and departure of all aircraft. Third Battalion, 6th Marines SOP was detailed, well-planned, and well-rehearsed. The PZ was selected by the battalion commander and the headquarters commandant was designated the PZCO.**

Staging Plan

Loads stand by at the PZ ready for the arrival of the aircraft. The staging plan restates the PZ organization, defines routes to the PZ, and provides instruction for linking up with the aircraft.

> **The staging plan was based on the loading plan and was covered in the battalion SOP for helicopterborne operations. It prescribed the arrival times of units at the PZ in the proper order for movement.**

APPENDIX A

SMALL-UNIT LEADER'S GUIDE TO PICKUP ZONE AND LANDING ZONE OPERATIONS

This appendix serves as a small unit (company and below) leader's guide for the safe, efficient, and tactically sound conduct of operations in and around PZs and LZs.

Selection and Marking

Small unit leaders should be proficient in selecting and marking PZs/LZs and in providing terminal guidance to aircraft. Tactical and technical considerations that impact selection of PZs/LZs are discussed in chapter 4.

The marking of PZs/LZs is as follows:

- During the day, a ground guide marks the PZ/LZ for the lead aircraft by holding an M-16A2 over his head, by displaying a folded VS-17 chest panel high, or by other identifiable means.
- The code letter Y (inverted Y) is used to mark the landing point of the lead aircraft at night (see fig. A-1). Chemical light sticks may be used to maintain light discipline.

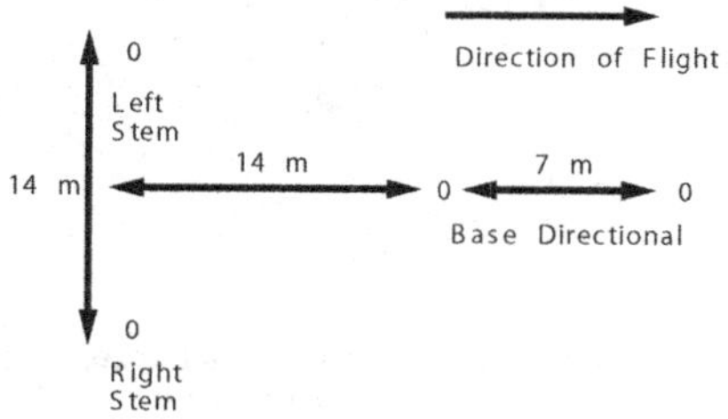

Figure A-1. Inverted Y.

- If more than one aircraft is landing in the same PZ/LZ, there will be an additional light for each aircraft. For observation, utility, and attack aircraft, each additional aircraft landing point is marked with a single light placed at the exact point that each aircraft is to land. For cargo aircraft, each additional landing point is marked with two lights and the two lights are placed 10 meters apart and aligned in the aircraft direction of flight.
- Obstacles include any obstruction to flight that might interfere with aircraft operation in the ground (trees, stumps, rocks) and cannot be reduced. During daylight, the aircrew is responsible for avoiding obstacles on the PZ/LZ. For night and limited visibility operations, all obstacles are marked with red lights. The following criteria is used in marking obstacles:
  - If the obstacle is on the aircraft approach route, both the near and far sides of the obstacle are marked.
  - If the obstacle is on the aircraft departure route, the near side of the obstacle is marked.
  - If the obstacle protrudes into the PZ or LZ, but is not on the flight route of the aircraft, the near side of the obstacle is marked.
  - Large obstacles on the approach route are marked by encircling the obstacle with red lights.

Control of Aircraft

Approaching aircraft are controlled by arm-and-hand signals to transmit terminal guidance for landing. The speed of arm movement indicates the desired speed of aircraft compliance with the signal. The signalman is positioned to the right front of the aircraft where he can best be seen by the pilot. Signalmen hold lighted batons or flashlights in each hand to give signals at night. If using flashlights, care must be taken to avoid blinding the pilot. Batons and flashlights will remain lighted at all times when signaling.

Assembly and Objective Areas

Prior to arrival of the aircraft, the PZ must be secured, PZ control party positioned, and the troops and equipment positioned in the unit assembly area.

Occupation of Unit Assembly Area

Unit leaders should accomplish the following:

- Maintain all-round security of assembly area.
- Maintain communications.
- Organize troops and equipment into loads and lifts in accordance with unit air movement plan.
- Conduct safety briefing and equipment check of troops.
- Establish priority of loading for each Marine and identify bump personnel.
- Brief on the location of the straggler control points.

Movement to the Occupation of Holding Area

Linkup guides from the PZ control party meet with designated units in the unit assembly area and coordinate movement of loads to a release point. As loads arrive at the release point, load guides move each load to its assigned load assembly area. (To reduce the number of personnel required, the same guide may be used to move the unit from the unit assembly area to the load assembly area.) If part of a larger helicopterborne assault, no more than three loads should be located in the load assembly area at one time. Noise and light discipline will be maintained throughout the entire movement in order to maintain the security of the PZ. Additionally, no personnel should be allowed on the PZ unless loading aircraft, rigging vehicles for sling load, or directed by PZ control. While remaining in load order, each Marine is assigned a security (firing) position by the heliteam commander or load leader. Each Marine is employed in the prone position, weapon at the ready, and facing outward (away from PZ) to provide immediate close-in security. An example of a large, one-sided PZ is depicted in figure A-2.

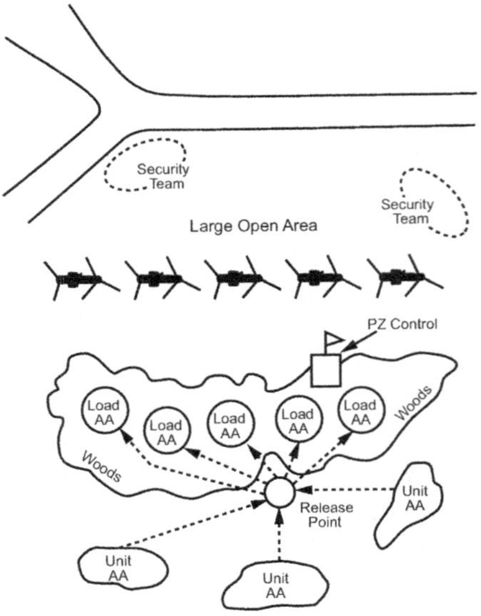

Figure A-2. Example of a Large, One-sided Pickup Zone.

An example of a small, two-sided PZ with unit and load assembly areas is depicted in figure A-3.

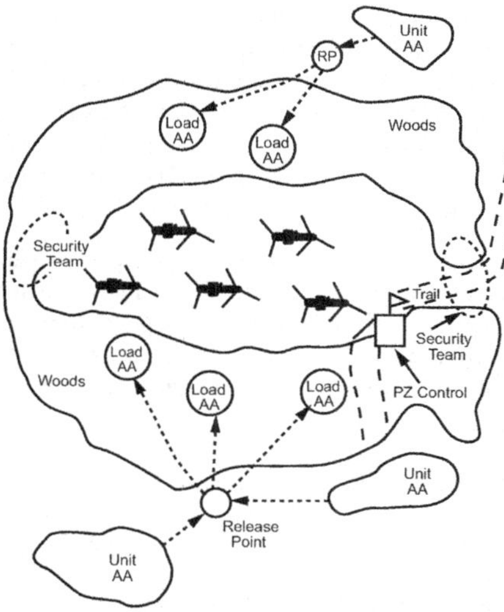

Figure A-3. Example of a Small, Two-sided Pickup Zone With Unit and Load Assembly Areas.

While in the load assembly area, units should adhere to the following for loading the aircraft:

- Maintain tactical integrity by keeping fire teams and squads intact.
- Maintain self-sufficiency by loading a weapon and its ammunition on the same aircraft.
- Ensure key Marines, weapons, and equipment are cross-loaded among aircraft to prevent the loss of control, or all of a particular asset, if an aircraft is lost.
- Ensure all troop gear is tied down and checked and that short antennas are placed in radio, folded down, and secured prior to loading.
- Ensure that squad and team leaders check the equipment of their Marines to ensure it is complete and operational.
- Ensure specific aircraft seats are assigned to each Marine.

PZ Closure

During platoon helicopterborne operations, the platoon sergeant is responsible for ensuring all personnel and equipment are loaded (clear the PZ) and security is maintained.

Single Lift

The platoon sergeant positions himself at the last aircraft and collects bumped Marines, if required. He will be the last person to load the aircraft. Once on the aircraft, the platoon sergeant uses the troop commander's radio handset to notify the crew chief/mission commander that all personnel and equipment are loaded. Close-in security is provided by the aircraft door gunners.

Multiple Lifts

The duties of the platoon sergeant during a multiple lift are the same as for the single lift. During a multiple lift, security teams maintain security of the PZ and are the last element to depart with the platoon sergeant. Depending on the initial location(s) of the security teams, repositioning closer to the PZ may be necessary. Whenever possible, the aircraft lands as close to the security team positions as possible to enhance security and minimize the movement required by the teams.

Helicopter Loading Sequence

The following helicopter loading sequence is used:

- The heliteam leader initiates movement once the aircraft has landed.

- The heliteam moves to the aircraft in file with the heliteam leader leading the file.
- The heliteam leader should—
  - Ensure that all personnel know which aircraft and which position to load.
  - Ensure that all personnel wear or carry packs on the aircraft.
  - Notify the crew chief when all heliteam members are on board and ready for liftoff.
- All personnel buckle up as soon as they are seated in their assigned seats. The heliteam leader always sit in the left front seat unless a platoon commander or company commander is on the same aircraft.
- The heliteam leader reports to the pilot and answers any questions the pilot may have, using the aircraft intercommunication (troop commander's) headset.

Landing Zone Operations

Just as there is a priority of work for defensive operations, there is a priority of actions upon landing in an LZ.

Unloading

Unloading the aircraft does not begin until directed by the crew chief or pilot. Once the aircraft has landed, personnel unbuckle seatbelts and exit aircraft as fast as possible with all equipment. Prior to leaving the aircraft, the heliteam leader obtains the landing directions from the pilot if not determined during the approach into the LZ; this aids in orientation to the LZ, particularly at night. Upon exiting the aircraft, the helicopter team moves to its designated location within the LZ.

Immediate Action on a Hot LZ

If the decision is made to use a hot LZ or contact is made upon landing, troops quickly dismount and move 15 to 20 meters away from the aircraft and immediately return the enemy's fire to enable the aircraft to depart the LZ.

If the contact is similar to a far ambush, troops will fire and maneuver off the LZ to the closest side offering cover and concealment. If troops are engaged from nearby enemy positions, they treat it as a near ambush by immediately returning fire. Marines who consider themselves in the kill zone may assault the enemy position(s) or attempt to get out of the kill zone. Marines not in the kill zone provide supporting fire to support the movement of Marines in the kill zone. The squad or platoon leader calls for fire support if it is available.

Once disengaged from the enemy force, the squad or platoon leader moves the unit to a covered and concealed position, accounts for personnel and equipment, and assesses the situation as to whether or not the unit can continue the mission.

Note

Expect assault aircraft gunners to return enemy fire when aircraft arrive and depart the LZ and while the aircraft is in the LZ.

Load Assembly Area in a Cold LZ

Upon unloading from the aircraft, the heliteam leader moves the load to its predetermined location using traveling overwatch movement techniques. All troops move at a fast pace to the nearest concealed position. Once at the concealed assembly point, the heliteam leader makes a quick count of personnel and equipment and then proceeds with the mission.

Duties of Key Personnel

To ensure that a helicopterborne operation is executed in an effective manner, key personnel are designated to perform specific duties.

Unit Leader

Platoon Commander

- Has overall responsibility for the helicopterborne operation. May act as the PZCO.

- Plans the operation.
- Briefs subordinate leaders.
- Issues OPORD.
- Conducts rehearsals.
- Rides in the AMC's aircraft to ensure better command, control, and communications.

Platoon Sergeant

- Sets up the PZ.
- Supervises marking of the PZ.
- Briefs all heliteam commanders.
- Supervises all activity in the PZ.
- PZ security.
- Movement of troops and equipment.
- Placement of loads and sling loads.
- Devises and disseminates the bump plan.
- Rides in the last aircraft for control purposes and ensures that the PZ is cleared.

Heliteam Leader

- Briefs his personnel on their respective tasks and positions inside the aircraft.
- Assigns respective areas of security to his personnel. Ensures that each Marine goes to his proper place.
- Supervises the loading of his helicopter team into the aircraft to ensure that all personnel assume assigned positions and have buckled their seatbelts.
- Keeps current on location by use of his map and communications with the aircraft crew during movement.
- Ensures, upon landing, that all personnel exit the aircraft quickly and move to designated positions within the LZ.

Pickup Zone Control Party

The PZ control party is responsible for the organization, control, and all coordinated operation in the PZ. A PZ control party for a platoon helicopterborne operation may organize as follows:

- PZ control officer is a rifle platoon commander.
- PZ control noncommissioned officer in charge is a platoon sergeant or guide.
- The radio operator has three radios: one radio monitors the aviation net for communication with the aircraft, another is used to communicate with the platoon's subordinate units, and a third operates on the company command net.
- There is one load linkup guide per helicopter team or load. His primary duties are to assist in linkup and movement of load assembly area. For platoon-sized helicopterborne operations, these guides should be selected from the same helicopter team squad they are assigned to.
- The lead aircraft signalman is responsible for visual landing guidance for the lead aircraft. The signalman can be selected from either the helicopter team or the squad that is loading on the lead aircraft.
- The hookup sending team is responsible for load preparation and rigging. The hookup sending team consists of a team supervisor/safety observer, an inside director, an outside director, a static discharge man, and two hookup men to hook up the load. A simple sling load may only require two Marines. During complex external lifts, Marines from the battalion HST may be provided to supervise platoon HST operations.

Helicopterborne Operations Communications

During establishment and operation of a platoon-sized PZ, communications must be maintained with aviation elements in order to control the aircraft. Communications are also required to report to the company headquarters and to control security teams. During air movement, radio listening silence is maintained on the company and platoon radio nets unless otherwise directed. While on board the aircraft, all leaders communicate with their troops using predetermined arm-and-hand signals or stating the information on a piece of paper or event map. Communication with the aircraft crew is accomplished by using the troop crew chief's handset.

Landing Zone Communications

Immediately after unloading the aircraft, radio operators check their radios to ensure they are configured as necessary. Radio communications on the LZ will be on the command frequency. Arm-and-hand signals and messengers are used to improve noise discipline.

Fire Support Communications

The rifle platoon makes requests for fire support through the rifle company headquarters.

Sling Load Hookup Operations

In small sling load hookup operations, company and below, six Marines are normally used as the ground crew in the PZ and LZ: a team supervisor/safety observer, an inside director, an outside director, a static discharge man, and two hookup men when needed by the platoon.

Static Discharge Equipment

The static electricity generated by helicopters during flight can be fatal to Marines conducting external loading and unloading operations. The static discharge grounding wand (National Stock Number [NSN] 1670-01-194-0926) protects the user from static electrical shock during helicopter external loading operations. It is important that all Marines conducting external load functions be trained in proper procedures and familiar with safety requirements. Marines should be trained in HST operations and external load procedures by landing support battalion, FSSG. The equipment required to conduct HST operations, including external loading, is drawn from the landing support company prior to attempting external loading and unloading. More information on HST operations is found in MCRP 4-11.3E.

Protective Equipment

All ground crew personnel will wear the following protective equipment:

- Helmet.
- Protective mask or dust goggles.
- Earplugs.
- Gloves.
- Utility shirt with sleeves rolled down.

Ground Crew Emergency Procedures

The hookup team will work on the right side of the load in order to move out to the right of the aircraft in case of emergencies. If an emergency occurs during a sling loading operation, the ground crew moves to the helicopter's right and the helicopter moves to its left. The signalman moves out of the helicopter's flight path by moving to the helicopter's right.

Safety Briefing

Prior to a helicopterborne operation, leaders within the unit chain of command give a safety briefing to all personnel. All leaders will enforce strict safety measures when working with helicopters. At a minimum, the safety briefing address the following issues:

- Identification tags and earplugs will be worn at all times when near or in an aircraft.
- Helmets, with chinstraps fastened, will be worn at all times.
- Helicopter safety measures for avoidance of tail rotors and proper loading and unloading procedures.
- M-16 rifles will be carried with the muzzle pointed downward, pistol grip forward, bolt closed, magazine in the weapon, and weapon on SAFE. Rounds will not be chambered. Bayonets will not be fixed.
- Hand grenades will be secured.

- Short antennas will be bent completely down and long antennas will be tied down when using radios in proximity of helicopters.
- Seatbelts will be fastened upon entering the helicopter and left buckled until the crew chief signals to exit the aircraft.
- In the event of a forced landing, all personnel will lean forward with their heads down until the aircraft comes to rest. No one will exit the aircraft until the main rotor has completely stopped.

APPENDIX B

AIR MISSION COMMANDER'S OR AIR OFFICER'S CHECKLIST

This list summarizes the essential items included in the planning phase of the helicopterborne operation by the air mission commander. The list is referred to throughout the planning process to ensure major item are not omitted.

1. Action Prior to Departure for Supporting Unit

Meet attack helicopter and reconnaissance representatives at pre-arranged site. Obtain briefing from designated helicopter unit operations officer that addresses support unit(s), mission, and planning data.

a. Support unit(s).

(1) Mission.

(2) Location.

(3) Contact officer.

(4) FM frequency.

(5) Call sign.

b. Mission.

(1) Requirements for aerial reconnaissance.

(a) Utility helicopters.

(b) Attack helicopters.

(c) Other assets.

(2) Special mission requests.

(3) Number of aircraft, by type, that are available for the operation.

(4) Utility, observation, cargo, and C2 helicopters required.

(5) Attack helicopters required.

c. Planning data for aircraft load for each type of aircraft.

(1) Number of troops _____; pounds of cargo _____.

(2) Number of reconnaissance available and time available.

(3) HST equipment available.

(4) Specific problem areas or requirements that may affect support of ground unit (FARP location and time of operation). (Estimated refueling time and refuel-rearm plan.)

(5) Obtain necessary equipment that will be required at or by the supported unit. (Special attention to HST external load requirements.)

(a) Aircraft or vehicle.

(b) Map, overlays, photographs.

(c) Radios, CEOI for exchange.

(d) Personal gear.

(e) Additional headsets for reconnaissance, if required.

(f) Slings, nets, or other HST equipment.

(6) Check with the HUC for special instructions.

2. Actions En Route

a. Establish and maintain communications.

b. Obtain status of fires and permission to enter area of operations.

3. Actions at Supporting Unit Location

a. Report to supported commander, S-3, or air officer.

b. Brief supported unit on number and type of aircraft available, aircraft load, and other essential information.

c. Obtain initial briefing on the following:

(1) Enemy situation.

(2) Friendly situation.

(3) Ground tactical plan (make map overlays).

d. Assist supported unit in planning.

(1) Movement to PZ for ground and aviation unit and control facilities.

(2) Loading.

(a) Location and selection of PZ.

(b) Special PZ marking procedures.

(c) Aircraft marking procedures.

(d) Landing formation and direction.

(e) Loads (HWSAT and HEALT).

1 Troops.

2 Cargo.

(f) Communications control procedures.

(g) PZ control (obtain call sign and frequency).

(h) Manifesting.

(i) Prioritization of bump by aircraft.

(j) PZ and lift-off times.

(k) GO/NO-GO criteria.

e. Air movement.

(1) The flight route provides guidance and information relative to flight times.

(2) Select en route formation that gives the most control and is least vulnerable to enemy interference; provide guidance for selection of PZ and LZ formation.

(3) Altitude and speed.

(4) Overwatch and security plan for attack helicopters.

(5) Fire support plan en route.

(6) Air movement plan.

(7) Reconnaissance HST support (finalize).

f. Landing.

(1) Touchdown times (in terms of H-hour).

(2) LZ location and designations and locations.

(3) Size and description.

(4) LZ marking and procedures.

(5) Landing directions.

(6) Landing formations.

(7) Traffic pattern for sequent lifts.

(8) Communications, control procedures, and use of reconnaissance.

g. LZ preparatory and suppressive fires.

(1) CAS (start time, duration, target and type of fuze, special instructions).

(2) Indirect fires (start time, duration, target and type of fuze, special instructions).

(3) Plan for attack helicopter unit's scheme of maneuver and plan for overwatch and security (start time, duration, special instruction, attack direction).

(4) Firing plan of debarking troops.

(5) Call signs and/or frequency signals for lifting and/or shifting support fires.

h. Refueling requirements.

(1) Location of FARP(s).

(2) Time required.

(3) Determine amount of fuel required to support mission

i. Aircraft maintenance.

(1) Downed aircraft procedures.

(2) Spare aircraft procedures.

4. <u>Actions Prior to Departure From Supported Unit</u>

a. Obtain copies of OPORD with overlays and annexes.

b. Confirm all times.

c. Last minute weather check.

(1) Mission procedures (delay increments).

(2) Alert procedures.

d. Debrief the commander.

5. <u>Actions Upon Return to Aviation Unit</u>

a. Inform unit commander.

b. Brief personnel, as appropriate, on all above information.

c. Maintain close liaison with support unit.

APPENDIX C

HELICOPTERBORNE UNIT COMMANDER'S CHECKLIST

This list is designed to summarize the essential items that should be included in the planning phase of a helicopterborne operation by the HUC. The list should be referred to throughout the planning process to ensure that major planning steps are not omitted.

1. Action Upon Receipt of Orders

   a. Analyze mission(s).

   b. Determine specified and implied task(s) and objective(s).

   c. Develop time schedule.

   d. Obtain aircraft availability information from mission commander and/or air officer.

   e. Issue warning order.

2. Ground Tactical Plan

   a. Choose, as appropriate, assault objectives.

   b. Designate LZ(s) available for use. Consider distances from LZ(s) to objective(s).

   c. Establish D-day and H-hour (time of assault).

   d. Identify special tasks required to accomplish mission.

   e. Means available to accomplish mission.

      (1) Organic troops (consider distance from present location to PZ).

      (2) Aviation resources to include attack helicopters (establish liaison with mission commander and obtain initial information relative to support requirements from ground unit to include FARP support).

      (3) Engineers.

      (4) Fire support.

         (a) CAS.

         (b) Artillery within range.

         (c) Other indirect fire weapons (mortar and naval gunfire).

         (d) Preparation fires for LZs (signals for lifting/shifting).

         (e) Flight corridors.

(f) Air defense suppression.

(5) Control measures needed.

(6) Subsequent operations (e.g., defense linkup, withdrawal) that may be conducted.

3. Intelligence Information and Requirements

a. Enemy locations to include air defense positions.

b. Commander's aerial reconnaissance of objective area (if practical).

c. Aerial reconnaissance

d. Sensor reports.

e. Terrain study.

f. Weather forecast.

g. Latest intelligence summary.

h. POW handling procedures.

i. Civilian control procedures.

j. PZ/LZ information.

k. Approach and retirement lane information.

4. Landing

a. Selection of primary and alternate LZ(s) (capacity).

b. LZ identification procedures for landing sites.

(1) Colored smoke.

(2) Panels.

(3) Flares.

(4) Lights.

c. Use of reconnaissance/S-2.

d. Landing formation(s).

e. Approach and departure directions.

f. LZ preparation fire to support landing plan and ground tactical plan.

(1) Use of CAS, air defense suppression.

(2) Use of indirect fire weapons.

(3) Use of EW.

g. Other fire support considerations.

(1) Shifting of fires.

(2) Lifting of fires.

(3) SEAD.

5. Air Movement

a. Flight routes (primary, alternate, return).

(1) Direction and distance to LZs.

(2) Checkpoints along route.

(3) Phase lines of used.

(4) Estimate time in route.

(5) Laagers (if used) to indicate location, mission, and security.

(6) Friendly air defense considerations.

(7) Enemy air defense intelligence.

b. Information to implement air movement.

(1) Units to be lifted.

(2) Number of types of lift helicopters allocated to each unit.

(3) Lift capability (maximum weight) for each aircraft type.

(4) Aviation units supporting unit.

(5) Lift off times.

(6) Routes.

(7) Unit LZs.

(8) L-hour (landing time of initial lift).

c. Alternate communications plan.

(1) FM.

(2) UHF.

(3) VHF.

(4) Visual/audio signals.

(5) Aerial radio relay.

6. Loading

a. PZ assignment by unit (primary, alternate) (bump/ straggler contingency plan).

b. Holding areas.

c. Routes from assembly areas to holding area to PZ(s).

d. Attack helicopter utilization (overwatch and security).

(1) En route to PZ.

(2) While lift aircraft are in PZ.

(3) En route to LZ.

(4) Reconnaissance of LZ; marking of LZ in absence of ITG.

7. Support Plans for Conduct of Helicopterborne Operations

a. Alternate plans and procedures due to weather (H-hour increment to delay operation).

b. Downed helicopter procedures.

(1) Crew and passenger duties.

(2) Aircraft disposition instruction.

c. Rally points.

d. Escape and evasion instructions.

e. Laager sites.

f. Rules of engagement.

g. Deception plans that will be used.

h. Spare aircraft available.

i. Reconnaissance (air-ground) that will be conducted.

j. Straggler control procedures.

k. Reporting (en route, liftoff, touchdown, intelligence, and contact).

l. Aircraft disposition after assault.

m. Medical support and evacuation procedures.

8. Actions That Must be Completed

a. Warning orders.

b. Liaison officer (receive and dispatch).

c. Briefing (time and place).

d. Preparation of OPORD.

e. Issue OPORD (time and place).

9. Logistics Requirements

a. Class V resupply.

b. Feeding plan.

c. Weather.

d. Casualty evaucation (call sign, frequency, location, and procedures).

e. Refueling (location of FARP, ammunition available).

10. Debriefing

a. Lessons learned.

(1) Ground units.

(2) Aviation units.

b. Actions taken for correction.

APPENDIX D

LANDING ZONE BRIEF

The landing zone brief is given prior to the transport helicopter landing in the LZ. LZ brief radio transmission are addressed by line number. Unknown or not applicable line numbers are referred to as negative. For example: line one-negative, line two-98632, line three-C3E, line four-negative, etc.

LANDING ZONE BRIEF				
1. MISSION NO.	____________			
2. LOCATION COOR/RAD/DME	____________			
3. UNIT CALL SIGN	____________			
4. FREQUENCY	PRI UHF	______	FM	______
	SEC UHF	______	FM	______
5. LZ MARKING	____________			
6. WIND DIRECTION/VELOCITY	______ / ______			
7. ELEVATION/SIZE	______ / ______			
8. OBSTACLES	____________			
9. FRIENDLY POSITIONS: DIRECTION/DISTANCE	______ / ______			
10. LAST FIRE RECEIVED: TIME/TYPE	______ / ______			
11. DIRECTION OF FIRE/DISTANCE	______ / ______			
12. CLEARANCE TO FIRE: DIRECTION/DISTANCE	______ / ______			
13. APPROACH/RETIREMENT (RECOMMENDED)	______ / ______			
14. PERSONNEL/EQUIPMENT	______ / ______			
15. OTHER	____________			

APPENDIX E

EXAMPLE OF AN ANNEX TO BATTALION SOPS FOR HELICOPTERBORNE OPERATIONS

1. General

a. Purpose. This annex prescribes the organization and procedures to be followed in the preparation and execution of helicopterborne operations. Only procedures peculiar to helicopterborne operations are included; otherwise, basic SOPs apply.

b. Application. Applies to all organic and supporting units under control of _____ Battalion, _____ Marines/// _____MEU, MEB, MEF.

2. Personnel

a. Strength, Records, and Reports

(1) Companies are organized into assault and rear echelons. Upon receipt of warning order, companies submit strength to S-1 and equipment availability status to S-4. S-1 and S-4 forward this information to the S-3 in order to determine flight requirements.

(2) Upon entering LZ, companies report personnel and equipment status to the battalion command post on the tactical net using the standard format.

b. Discipline and Order

(1) S-1 establishes a straggler control point on each PZ in coordination with the S-3 and PZCO. All units will have a representative located at the straggler control point on their PZ(s). Bumped personnel are reported to the S-1 and/or PZCO by straggler control personnel for consolidation and rescheduling into appropriate LZ(s).

(2) Straggler control becomes the company's responsibility upon landing.

(3) Personnel landed in other than assigned LZ are to report to the on-site unit commander (representative) immediately. Personnel are attached to that unit until ordered to return to parent unit by this headquarters. Gaining unit reports attached personnel to the S-1 by their number and parent unit. Do not include attached personnel in unit strength reports.

c. Prisoner of War Evacuation. POWs are immediately reported to the S-3, who issues evacuation instructions. Units detaining POWs indicate available PZ location for pickup by air in their initial reports. The S-2 determines whether to evacuate through battalion or direct to higher headquarters.

d. Casualty Evacuation. Report all casualties for casualty evacuation by priority code.

(1) Casualty evacuation requests follow standard format for casualties and are classified as routine, priority, or urgent. Radio frequency of supporting casualty evacuation units are included in each OPORD. When casualty evacuation aircraft are not available and immediate casualty evacuation is required, make maximum use of empty lift helicopters departing LZ. Do not interrupt air loading operations—use last helicopters in the flight.

(2) The S-1 is responsible for providing CEOI, LZs, and flight route overlay to supporting casualty evacuation units.

3. Intelligence

a. Weather

(1) Battalion S-2 obtains and disseminates the following:

- Long-range forecast immediately after receipt of mission.
- Short-range forecasts up to H-2.

(2) Command weather reconnaissance 1 hour prior to lift-off is coordinated among the commander, S-2, and air mission commander.

(3) Operations are executed only on order of this headquarters when weather is below half-mile visibility and 100-foot ceiling.

b. Terrain

(1) Maximize use of command aerial reconnaissance down to company commanders, consistent with aviation resources, available time, and tactical situation.

(2) Use sand table briefing techniques when possible in conjunction with maps and aerial photographs.

(3) Issue maps immediately after receipt of warning order, if available. If not available, they are issued on receipt from higher headquarters.

(4) Aerial photographs are made available upon receipt. The S-2 processes all requests (conserving assets, as appropriate). Priority is given to answering priority intelligence requirements and support of the assault echelon.

c. Escape and Evasion

(1) Personnel in aircraft forced to land behind enemy lines en route to the objective will:

- Establish immediate security in vicinity of downed aircraft.
- Remain in that location using aircraft radios to contact recovery aircraft.
- Mark and clear suitable landing points for recovery and withdrawal helicopters.

(2) The senior ground force individual assumes responsibility for organization and conduct of security until recovery is executed.

(3) If the above is not possible due to enemy pressure, evade capture and attempt to join friendly units by infiltration. Personnel attempt to escape and evade back along flight route because maximum rescue efforts are directed along flight routes, with particular emphasis at

checkpoints. Continuous attempts will be made to locate suitable LZ/LP for withdrawal by helicopter(s). The wounded are evacuated with infiltrating personnel. The dead are concealed and stripped of weapons, ammunition, and items of intelligence value. The locations of the dead left behind are recorded.

(4) If enemy pressure becomes a threat to downed aircraft personnel, the senior ground force individual take steps to secure or destroy classified or sensitive items. Aircraft destruction is only on order of this headquarters if capture is not imminent. If contact with this headquarters cannot be made, the senior individual on the ground attempts to prevent capture by enemy.

4. Operations

a. Planning Phase

(1) Except when accomplished by higher headquarters, this headquarters prepares plans in coordination with the supporting air mission commander.

(2) Plans continue to be refined until executed. All operational information is given to subordinate commanders as soon as determined, particularly—

- The size and composition of the force required to execute the mission.
- Allocation of assault and logistical aircraft, based on allowable cargo load provided by the mission commander.
- Designated PZs and helicopter PZ formations. Designated flight routes, LZs, and LZ helicopter formation.

(3) Coordination between the supported and supporting commanders must include, at a minimum—

- Enemy and friendly situation.
- Mission.
- Fire support plan.
- Abort and alternate plans.
- Weather (including minimums and delays).
- Type, number, and aircraft load of helicopters.
- Helicopter formations in PZ and LZ.
- Air movement information relative to timing for operations.
- Communication (primary and alternate frequencies and plans).
- Location and call sign of second in command.
- Required command reconnaissance by the helicopterborne commander and supporting aviation commanders.
- Time synchronization requirements.
- TRAP procedures.

(4) Operations Security. OPSEC is emphasized in each phase of a helicopterborne operation. The objective is to conceal the capabilities and intentions of the helicopterborne force. Four, general OPSEC measures are considered for every operation: deception, signal security, physical security, and information security. The S-2 provides intelligence collection of threat data. The S-3 ensures that the staff and subordinate commanders are aware of OPSEC measures

to be employed to counter the threat. Emphasis is placed on maintaining the elements of surprise and security. Additionally, all supporting elements must be aware of the necessity of maintaining a high degree of operational security. At a minimum, the commander, supporting commanders, and subordinate commanders employ the following techniques:

- Deception—
  - Camouflage vehicles, equipment, and personnel.
  - Overflights of other LZs (if aircraft are available and enemy situation does not preclude).
  - Insertion at night or during other periods of reduced visibility.
  - Noise and light discipline.
  - Dummy larger sites for aircraft.
  - Reconnaissance overflights of several objectives.
- Signal security—
  - Communications security techniques.
  - Radio listening silence.
  - Use of hand and arm signals (on the ground).
  - Use of low power and secure mode on radios.
- Physical security—
  - Use of security forces at LZ and PZ.
  - Use of wires, mines, barriers, and security troops at aircraft larger sites and troop assembly areas.
  - Use of reconnaissance unit to secure LZ, if possible.
- Information security—
  - Counterintelligence.
  - Strict control of all operational information.
  - Release information only to those with a need to know.
  - Last minute release of attack time (objective) and force composition.

b. Landing Phase

(1) The aircraft commander notifies each helicopter team leader of any changes to the order, any change in LZ(s) and/or direction of landing, and when the helicopter is over the release point. The helicopter team leader then informs his personnel of any changes and alerts them to prepare to unload.

(2) Passengers may not move in the aircraft until clearance has been obtained from the helicopter team leader via the helicopter crew chief. After the helicopter team leader gives the clearance signal, troops and equipment are unloaded as rapidly as possible.

(3) After all troops and cargo have been unloaded from the aircraft, the crew chief checks the helicopter and signals the helicopter team leader that the cabin is empty. Departure from the aircraft is executed rapidly in the direction prescribed by the battle drill.

(4) The helicopter team leader ensures that members of his helicopter team clear the LZ in a safe, expeditious manner. This prevents exposing personnel to unnecessary danger and prevents any delay in lift-off and landing of subsequent helicopters.

(5) Individual weapons are fired only upon order during offloading unless enemy contact is made or if planned as part of the overall fire plan.

(6) If there is no enemy contact on the LZ, actions are as follows:

- Move each helicopter load to the nearest covered and concealed position in direction of the assembly area.
- Establish LZ security for succeeding lifts (if applicable).
- Assemble, organize, and account for all personnel.
- Report.

(7) If enemy contact is made on the LZ, actions are as follows:

- Return fire immediately, upon offloading, with all available firepower to gain fire superiority.
- Fight by helicopter loads, using fire and movement, until platoon or company can be formed (according to battle drill).
- Request and coordinate fire support.
- Secure the LZ for succeeding lifts.
- Report.

(8) Keep the commander informed during all actions.

c. Air Movement Phase

(1) Maintain radio silence to maximum extent possible. Inability to comply with specific control times are reported as prescribed in OPORD.

(2) Troop leader remains oriented by continuous map-terrain comparisons.

d. Loading Plan

(1) The PZ is designated by this headquarters.

(2) The air officer arrives prior to the helicopter flight and reports to the PZCO for last-minute briefing and coordination. The air officer notifies the air mission commander of any changes.

(3) Serials are organized to support the ground tactical plan.

(4) Helicopters land in the PZ(s) in the specified formation. Unit leaders brief troops on the helicopter formation prior to arrive of helicopters at PZ.

(5) Helicopters should land as close to their estimated time of arrival as possible to reduce time-on-ground before loading.

(6) During a battalion move, the battalion XO, or designated representative, acts as the PZCO and the headquarters commandant acts as the LZ control officer. The company XO acts as PZCO during company-sized operations and as unit PZCO during battalion-sized operations. PZCOs are responsible for developing and disseminating the PZ control plan. The PZCO maintains contact with the AMC on a designated radio frequency. Each unit to be moved will have radio contact with the PZCO 15 minutes prior to aircraft arrival. Units must be prepared to alter loads based on change of helicopter availability or change in allowable cargo load. Within

each company, platoon, and squad, a priority of loading must be established. Priority of aircraft loads and personnel on each aircraft to be bumped will be designated. Bumped personnel report to the straggler control point immediately.

(7) The supporting aviation unit assists in planning for the execution of loading by providing technical advice and supervision.

(8) The heliteam leader supervises helicopter loading.

(9) Cargo or equipment to be transported externally is secured in cargo nets or on pallets for sling loading under helicopters. Hookup of these slings is accomplished by the HST in the PZ.

(10) Preparations for individuals are as follows:

- Fasten helmet chinstraps.
- Collapse bipods on M-240s and M-16s.
- Tie down loose equipment.
- Place all weapons in a condition 3 status.
- Unfix bayonets (if fixed).
- Radio operators use short whip antennas only and depress antennas to avoid breakage and to reduce the safety hazard. When directed, they check communications with the tactical operations center, ensuring that the radio remains on during flight. They will have a minimum to two extra batteries for each radio.

(11) An accurate list for each aircraft load by name, grade, and unit is furnished to the battalion S-1 through the unit officer in charge of loading.

(12) The following sequence should be followed during the loading phase:

- Secure PZ.
- Approach aircraft only after it has landed.
- Do not load until directed by PZ control personnel.
- Load at double time.
- Move to the aircraft and load as directed by the heliteam leader.

(13) When loading personnel and cargo into a helicopter, the heliteam leader ensures that the following is accomplished:

- All safety measures prescribed for movement in and about the helicopter are observed.
- All personnel approach the helicopter in the prescribed manner.
- Personnel are aware of and avoid the tail rotor and engine exhaust outlets.
- All personnel and equipment stay below the arc of the top rotor and load on the opposite side of the tail rotor. Personnel should be especially watchful when loading on the slope of a hill; therefore, approach and depart a helicopter on downslope side. The aircrew indicates which side to enter/exit the helicopter.

(14) Briefing on emergencies are conducted by an aviation representative prior to loading, as appropriate.

(15) Upon loading, the helicopter team commander provides the gross weight of the load (personnel + all equipment) to the pilot in command or the crew chief.

(16) After all equipment and personnel have been loaded, the helicopter team commander, in coordination with the aircraft crew chief, determines that—

- Equipment and cargo are in the proper places.
- Cargo or equipment is properly secured.
- Each Marine is seated and his safety belt fastened.
- Weapons are placed between legs; muzzle down (except on the UH-1N which is boarded muzzle up).

(17) When the helicopter team commander has checked to ensure that all cargo and personnel are accounted for, he notifies the aircraft commander or crew chief. The crew chief ensures that all personnel and equipment are properly secured for flight.

(18) During flight, the pilot commands the aircraft. The helicopter team leader ensures the following is accomplished:

- Cargo lashing (if applicable) is checked to determine that cargo is properly secured.
- Troops keep belts secure and do not smoke or sleep during flight.
- Troops stay seated and do not move around without authorization.

(19) In the event that more than one lift is required, the PZCO remains until the last lift to ensure control and continuous communication.

(20) General aircraft load planning requires that—

- All units develop general load plans to facilitate movement on short notice.
- The necessary equipment for aircraft loading and movement (nets, slings, and clevises) is kept on hand.
- Battalion personnel are organized and trained in loading equipment (including sling loads).
- Vehicles and major equipment are prepared at all times to facilitate airlift operations. Vehicles and major equipment to be transported into objective area are reported with strength figures.

e. Subsequent Operations

(1) Withdrawal by Air. Withdrawal from an area of operations requires thorough planning, close coordination, and controlled execution. The following are considered important for any withdrawal by air:

- Primary and alternate PZs and flight routes must be planned.
- Defensive concentrations must be planned around the PZ. The security force protects the loading force and returns fire if engaged. When the last elements are ready to load, the security force calls in required fires to cover withdrawal and uses their own fire to cover their loading.
- All around security until the first helicopter is on the ground (never assemble too early) must be maintained.

- Loads must be planned so that a force is capable of defending itself until the last lift (never leave less than a platoon size force). The platoon leader/sergeant or squad leader with a radio is the last man out of a PZ and reports to his commander that the PZ is clear of all personnel and equipment and immediately notifies the pilot of the helicopter he boards. If possible, plan for at least two extra helicopters to go into the PZ to lift out the last unit.
- The attack helicopter unit will be in direct communication with the HUC.

(2) Displacement of Command Post

(a) The quartering party is composed of S-1 or headquarters commandant, communications officer or representative, communications personnel, security element, and other necessary personnel, and they select the location of the command post.

(b) The C2 helicopter is used as the main command post during movement.

(c) Quartering party duties upon landing include—

- Laying out the new command post.
- Notifying the old command post when the new command post is ready for operation.
- Ensuring timely and orderly arrival and positioning of other command post elements.
- Opening the new command post. The officer in charge notifies the commander or S-3 when the old command post has closed and when the staff is operational in the new location.
- Providing controlling responsibilities. A C2 helicopter is used as the tactical command post to control and direct subordinate elements during air movement. The old command post is responsible for the dissemination of information and reports to higher and adjacent headquarters until that function is formally passed to the new command post.

(3) Passive Security of Aircraft in Unit Areas. The security of supporting aviation is the responsibility of the unit commander in whose area they are laagered or as designated by the headquarters.

(a) Laagers (occupancy, 1 to 36 hours) have the following characteristics:

- Select proper terrain for laagers where access by enemy ground forces is difficult (e.g., laagers surrounded by water or swamps).
- Site aircraft to blend with terrain and vegetation (e.g., locate parking areas in shadows, near trees).
- Park aircraft in laagers so that attack helicopters can provide security along avenues of approach. Lift of aircraft, if attacked by enemy, is the responsibility of the ACE commander.
- Utilize troops in or near the laagers to provide perimeter security. Helicopter units augment security.

(b) Semipermanent facilities (occupancy, 1 to 7 weeks) have the following characteristics:

- Use camouflage nets and natural materials to provide concealment.
- Provide perimeter troop security around airfields and helipads.
- Construct individual and helicopter bunkers and continue progressive improvement as time permits.

5. Logistics

a. Supply

(1) Accompanying Supplies—All Classes. Prescribed supplies are established by the headquarters for each helicopterborne operation.

(a) Class I. Each Marine carries three combat ration meals to be eaten on order.

(b) Class II and IV. Units take one day's supply of required combat essential expendables.

(c) Class III. Vehicle fuel tanks are filled three-fourths full and gas cans are filled to the weld. Units take one day's supply of oil and lubricants on vehicles.

(d) Class V. Units maintain basic load at all times. Available supply rates and priority of delivery as specified in OPORD.

(e) Class IX. Units take combat essential prescribed load list.

(f) Water. Marines carry two full canteens and one bottle of water purification tablets.

(2) All classes of supply are delivered using unit distribution.

(3) Routine, planned supplies are prepackaged to the maximum extent possible by the S-4.

(4) Emergency resupply containing ammunition, water, rations, and medical supplies are prepackaged by the S-4 and are ready for delivery as required.

b. Salvage

(1) Expedite recovery of aerial delivery containers, cargo nets, and pallets; commanders guard against damage, destruction, or loss.

(2) Units in objective area establish salvage collecting points when appropriate and practical.

(3) Salvage is reported to this headquarters for disposition instructions.

c. Captured Material. Captured material may be used on approval of this headquarters. Evacuation of captured material is accomplished, as the situation allows, though S-4 channels.

d. Medical Support

(1) Casualty evacuation of patients, until linkup or withdrawal, will be by air.

(2) Aid station location is normally in the battalion's rear.

(3) Requests for casualty evacuation within the helicopterborne operations area is made to the medical organization on the casualty evacuation frequency or the administrative logistic net.

(4) POW casualties needing medical treatment are evacuated through medical channels.

(5) Hospital locations are announced for each operation.

e. Transportation and Troop Movement

(1) Vehicular

(a) Allocation of accompanying organic transport is made by this headquarters.

(b) Captured vehicles are used to the maximum to meet transportation requirements.

(2) Aircraft. Allocation of supporting aircraft is made by this headquarters.

6. Visual and Sound Signals

a. Use visual and sound signals as required and prescribed by CEOI and unit SOP and as modified by battalion OPORD.

b. Subordinate units employ only those pyrotechnics specifically authorized by OPORD or CEOI.

7. Electronic Warfare

a. Radio stations will not attempt to enter, jam, or otherwise interfere with unknown radio nets without prior approval of this headquarters.

b. Report (by a secure means) jamming or attempts to enter nets by unknown stations to the communications officer without delay. Give time, radio frequency, type of jamming, signal strength, readability, and identity (if obtainable) of interfering station.

Appendix F

Formats for the Five Basic Plans

Successful use of helicopters requires a careful analysis of METT-T and detailed, precise reverse planning. The formats provided in this appendix provide a guide for the development of the five basic plans that comprise a helicopberborne operation.

GROUND TACTICAL PLAN

All planning evolves around this plan. The plan specifies actions in the objective area that ultimately accomplish the mission. The information listed below provides a guide that can be used to establish the plan.

1. MAGTF Commander's Mission and Intent.

2. GCE Commander's Mission and Intent.

3. HUC Mission and Intent.

4. Forced Time Schedule.

   a. Time that assault elements land (L-hour).

   b. Reverse planning sequence.

Time	Event
____________	L-hour
____________	1st assault wave(s) arrive in PZ
____________	Assault element arrives in assembly area
____________	Assault element arrives in holding area
____________	Assault element arrives in PZ
____________	Issue warning order
____________	Intelligence (S-2) brief
____________	Commander's guidance
____________	Staff briefs
____________	Operations order
____________	Mission brief

5. Actions Required in the Objective Area.

Secure LZ __________ ____________________.
(name/coord) (name/coord)(name/coord)

Establish LZ control at above LZ(s).

Secure Objective(s)______________________________.
(name/coord) (name/coord)(name/coord)

Perform the following actions upon securing objective(s):

____

____

____

6. <u>Mission Assigned to Subordinate Units</u>.

Co __________________________

Co __________________________

Co __________________________

Co __________________________

Attachments with assault elements:

Follow on attachments:

7. <u>Coordinating Instructions</u>:

LANDING PLAN

The landing plan must support the ground tactical plan. The plan sequences elements into the area of operations so that units arrive at locations and times prepared to execute the ground tactical plan. The information listed below provides a guide that can be used to establish the plan.

1. Size and location of primary and alternate LZ(s):

Primary LZ

Size ____________ **Location** ____________

Size ____________ **Location** ____________

Size ____________ **Location** ____________

Size ____________ **Location** ____________

Alternate LZ

Size ____________ **Location** ____________

Size ____________ **Location** ____________

Size ____________ **Location** ____________

Size ____________ **Location** ____________

2. Known and suspected enemy locations in and around the LZ:

Size ____________ **Location** ____________

Size ____________ **Location** ____________

Size ____________ **Location** ____________

Size ____________ **Location** ____________

3. Unit tactical integrity and spread loading:

- Squads in one aircraft.
- Platoons in one wave.
- Key leaders NOT loaded on the same aircraft.
- Crew-served weapons and crews sufficiently spread loaded.

4. All members briefed and oriented to the landing:

- Briefed on actions at the LZ.
- Briefed on actions to secure the LZ.

5. Task organization:

- For landing.
- Subsequent to landing.

6. Determine who decides to switch to an alternate LZ:

- HUC.
- S-1.
- AMC.
- Assault flight leader.

7. Factors in deciding to switch to alternate LZ(s):

- LZ too hot.
- Downed aircraft in LZ.
- Escort warning of ambush.
- Other.

8. Plan for supporting fires:

- Planned fires for air movement .
- Planned fires for landing.
- Preparation fires in LZ.
- Preparation fires near LZ:
  - Distant preparation fires
  - Preplanned fires in and around LZ
  - On-call fires
- Planned fires subsequent to landing.

9. Plans for casualty evacuation:

- Air.
- Ground.

10. Plans for resupply:

- Air.
- Ground.

AIR MOVEMENT PLAN

The air movement plan is based on the ground tactical and landing plans. The plan specifies the air movement schedule and provides instructions for the air movement of troops, equipment, and supplies from PZs to LZs. The following information provides a guide that can be used to establish the plan.

1. Tentative flight routes are selected by the AFL. The HUC's S-2 studies the mutes and makes recommendations. The HUC's S-3 closely notes checkpoints and control features.

2. The air movement schedule is developed to accomplish the landing plan. The air movement schedule is provided by the helicopter transport commander. The HUC studies the schedule and makes recommendations.

3. Air speed, flight altitudes, and aircraft formations are determined by the AFL.

4. Escort of transport helicopters and air fire support during air movement is determined by the air commander.

5. Aircraft availability information is provided to the HUC

CH-46E ____________

CH-53D____________

CH-53E____________

UH-IN ____________

AH-lW ____________
fixed-wing available:

6. The wave allocation of transport helicopters is determined by the HUC.

	1WV	2WV	3WV	4WV
CH-46E	________	________	________	________
CH-53D	________	________	________	________
CH-53E	________	________	________	________
UH-1N	________	________	________	________

7. Wave allocation of escort aircraft is determined by the air commander

	WV	2WV	3WV	4WV
AH-lW	________	________	________	________

8. Air departure points from a start point in the sky to the LZ are determined by the air commander.

9. Loading times are determined by the air commander.

WavePZ Load Time

1 ____________

2 ____________

3 ____________

4 ____________

10. Lift off times are determined by the air commander.

WaveLift Off Time

1 ____________

2 ____________

3 ____________

4 ____________

LOADING PLAN

The loading plan is based on the air movement plan. It ensures that Marines, equipment, and supplies are loaded on the correct aircraft. Helicopter loads are also placed in priority to establish a hump plan. The following information provides a guide that can be used to establish the plan.

1. Refer to paragraphs 5 and 6 of the guide for the air movement plan.

   a. Review the total number (by type) of transport aircraft available.

   b. Review the number of aircraft by type allocated to each wave.

2. Determine which personnel, weapons, and equipment will be loaded on each aircraft.

   a. Maintain unit integrity.

   b. Spread load key personnel, weapons, and equipment.

3. Determine if the preparation of a written document is necessary. An informal document lists the personnel, key weapon, and equipment by aircraft. A formal document includes a HWSAT.

4. Establish a bump plan so that essential personnel and equipment are NOT unnecessarily delayed in case of aircraft complications.

   a. The plan defines who (by name) gets off each aircraft first, second, third, etc., in the event the aircraft cannot carry a full load.

Example of an Individual Bump Plan		
Aircraft	**Unload**	**Sequence**
101-1	Off 1st	Johnson
	2	Jones
	3	Smith
	4	Howard
	5	Stevens
	6	Britt
	7	Randall
	8	Bump entire load

b. The bump plan also defies when each aircraft load will subsequently be loaded in the event an aircraft cannot fly.

Example of a Load Bump Plan	
Aircraft	**Reload Plan if Bumped**
101-1	Next available CH-46
101-2	Next available CH-46 after 101-1
101-3	Next available CH-46 after 101-2
101-4	1st available CH-46 in 2d wave
101-5	1st available CH-53D or CH-53E

5. The ground commander designates unit loading sites.

6. The ground commander establishes the plan and procedure for controlling the arrival, loading, and departure of all aircraft.

7. The ground commander designates a PZCO.

STAGING PLAN

The staging plan establishes the specific sequence, loads, ground routes, guides, and times from the assembly area to the holding area and from the holding area to the PZ. The example listed below provides a guide that can be used to establish the plan.

Sequence	Load	Route	Guide	Depart Hold AR	ARVPZ
1	101-1	A	Lt Jones	0500	0515
2	101-2	B	SSgt Brown	0500	0515
3	101-3	C	Sgt Smith	0500	0515
4	102-1	A	Lt Wells	0515	0530

APPENDIX G

HELICOPTERBORNE TRAINING

Helicopterborne training must be integrated into unit programs on a routine basis to develop capability at each level from squad through battalion. Commanders are responsible for their unit's helicopterborne training. The objective is for units to conduct helicopterborne operations with speed, precision, and confidence. Infantry units, as well as other combat, support, and CSS units, should routinely receive such training.

Small-Unit Training

Standard, infantry small-unit tactics and techniques are the basis for the ground phase of helicopterborne operations. The commander ensures that all units are proficient in these tactics, and then combines this training with training that is specific to helicopterborne operations: staging, loading, air movement, landing, and unloading. The commander emphasizes the rapid loading and unloading of aircraft, as well as quickly organizing maneuver elements in the LZ to take advantage of the speed and mobility of helicopterborne operations.

The commander trains small-unit leaders to operate independent of their parent organization in order to accomplish their part of the overall mission. Additionally, small-unit leaders must be able to take charge in the absence of their seniors. The speed and complex nature of helicopterborne operations dictates the use of SOPs and battle drills.

Mobility

The commander trains his units to travel light, consistent with the mission, taking only necessary equipment and supplies.

Helicopter Egress Training

Marine Corps Underwater and Immediate Passenger Helicopter Aircrew Breathing Device (IPHABD) Familiarization Program. This program focuses on the use of the SRU-40/P IPHABD and trains them on the SWET (Shallow Water Egress Trainer). All four rides in the SWET are followed by familiarization with all the survival gear (rafts, survival strokes, treading water) and then four rides in the dunker (two with the IPHABD).

SOP and Training

Procedures for conducting helicopterborne operations are included in unit SOPs. While SOPs include routine actions that personnel might have to complete during an operation, they must also include procedures for downed aircraft, bump plans, or other conditions that may occur during the conduct of operations. The unit's training program ensures that personnel are familiar with and proficient in the procedures contained in the SOP. Also during training, the information in the SOP is evaluated for completeness, simplicity, and applicability and procedures are refined as necessary.

Land Navigation

Land navigation proficiency by all leaders is critical to success. Leaders must learn to locate positions, navigate to specific points, and use the terrain to their advantage.

Artillery Support

Supporting artillery units train with the maneuver unit. They become familiar with the maneuver unit's SOP and teach selected maneuver unit personnel how to plan for, employ, call for, and adjust artillery and mortar fires. This joint training ensures mutual understanding of operational requirements, capabilities, and limitations. If possible, the same fire support units support a particular maneuver unit for each operation.

To support helicopterborne operations, the artillery must be proficient in sling loading operations and the planning required to execute PZ/LZ operations. This planning requires artillery leaders to coordinate closely with both the maneuver unit that controls the lift assets and the aviation units involved. HST and external load/unload techniques require frequent training for hookup teams, helicopter crews, zone control personnel, and communicators. A detailed discussion of external loading and unloading procedures is contained in MCRP 4-11.3E.

Aviation Units

Aviation unit commanders assist ground unit commanders in the development of training in the technical aspects of combined aviation and ground unit training. They also ensure that their units are technically proficient. A working relationship between the maneuver and aviation units is maintained whenever possible.

Infantry and Aviation

Ground and aviation units must train together and completely understand the MAGTF concept. They must train in all types of weather and visibility. As a result of their joint training, they refine and develop compatible SOPs.

Developing Helicopterborne Training Programs

A training program for helicopterborne operations should include the critical individual and collective skills necessary to accomplish the warfighting mission. Unit training should identify weaknesses and train to correct the weaknesses.

Conduct of Training

Helicopterborne training begins by familiarizing individuals in aircraft procedures; this includes loading and unloading, crash procedures, and aircraft safety. Proficiency in these procedures provides a foundation for collective training of ground and aviation units.

Collective training should include battle drills on loading and unloading, as well as organizing into combat formations on the LZ. This training allow units to maximize the speed and mobility of helicopterborne operations.

Use of Mockups

Constraints on helicopter flight hours limit the amount of flight time available for training. Therefore, much of the individual and small-unit training has to be accomplished using aircraft mockups. Plywood and other materials can be used to build the mockups, which are relatively inexpensive. Mockups can be used to train individuals on how to approach a helicopter, how to get on it, and how to get off of it. Helicopterborne battle drills can be taught by using mockups. Combat support Marines can be trained to load weapons, equipment, supplies, and ammunition on helicopters by practicing on mockups.

If the unit has a local training area of adequate size, several mockups can be used to practice battle drills to include the way the unit should offload aircraft in the LZ. The mockups can be placed in different patterns to simulate different landing formations. Three CH-46 mockups should be sufficient for platoon training.

Individual and Unit Training

The following subjects should be included in appropriate phases of individual and unit training:

- Ground units:
  - Subjects required to obtain proficiency in ground skills and tactics.
  - SOP battle drills.
  - Physical and psychological preparedness.
  - Methods and procedures for control and guidance of aircraft.
  - Safety procedures in and around aircraft.
  - Control and adjustment of supporting fires.
  - Subjects required to obtain proficiency in preparing internal and external aircraft loads.
  - Practical experience in land and aerial navigation.
  - Employment of attack helicopter units.
  - Helicopter team commander's duties.
  - Helicopter rope suspension techniques.
  - Downed aircraft procedures.
  - LZ/PZ selection.
  - LZ/PZ control.
  - Combat support and CSS requirements and techniques.
- Aviation units:
  - Operations planning.
  - Terrain-flying techniques and navigation.
  - Formation flying.
  - Marginal weather and reduced-visibility flying techniques.
  - Camouflage and security of aircraft.
  - Employment of aerial weapon systems.
  - Aircraft maintenance in a combat field environment.
  - Unit control of aircraft and air traffic.
  - ITG procedures and techniques.
  - Flight operations in confined areas with maximum loads.
  - Operations with external loads.
  - Aerial reconnaissance and security techniques.
  - Battle drills.
- Subjects common to aviation and ground units:
  - Threat organizations and doctrine.
  - Recognition of threat vehicles and antiaircraft weapons and knowledge of their capabilities.
  - Conduct of liaison and coordination.
  - Forward refueling techniques.
  - Training in defense against NBC weapons.
  - Signal security, discipline, and electronic attack.
  - Casualty evacuation procedures.
  - Procedures for aerial resupply.
  - Training in helicopterborne SOPs.

Preparation

Training time and resources must be used efficiently. Each element of the unit should be prepared to do its part before joining support units for combined exercises. Squad and platoons should be trained in the following:

- Helicopterborne battle drill.
- Preparation of internal and external loads.

Staff Training

Staffs of ground and aviation units must be trained in planning and conducting helicopterborne operations with emphasis on the following:

- Capabilities and limitations of helicopterborne operations.
- Command and staff relationships in the MAGTF.
- Development of plans using the reverse planning sequence.
- The MAGTF rapid planning process.
- Fire support means and control and fire support planning for helicopterborne operations.
- Logistical procedures and requirements for helicopterborne operations.
- Preparation of sequenced ground and air movement plans.

APPENDIX H

MARINE CORPS HELICOPTER CHARACTERISTICS

To efficiently load an HTF aboard helicopters, ground commanders and staffs must know the exact composition of the force, the essential characteristics of the types of helicopters to be used for the operation, and the methods of computing aircraft requirements.

Maximum aircraft loads are affected by altitude and temperature and will differ widely according to topography and climate conditions common to specific zones or areas of military operations. Loads will further vary based on the location of, approaches to, and exits from LZs; pilot proficiency; aviation unit SOP; type of engine in the aircraft; and age of both aircraft and aircraft engine. Therefore, two identical aircraft may not be able to pick up and carry identical loads.

This appendix discusses the general characteristics of Marine Corps helicopters. Refer to the NWP 3-22.5 series for detailed helicopter information, technical data, and guidance for computing aircraft requirements. Also, see JP 3-02.1, *Joint Doctrine for Landing Force Operations*, for information and examples of detailed air loading and air movement forms. The same air movement forms common to amphibious operations can be used for subsequent operations ashore when such movement documentation is essential in planning and operations.

Aircraft Availability

Aircraft availability is the overriding consideration in helicopterborne operations. It is directly influenced by the adequacy and efficiency of maintenance and supply activities and aircraft utilization and scheduling procedures, as well as by the distance of support units from the operating units.

Both the support and supported commanders should be aware that everyday use, over an extended period, of all available aircraft results in a reduced mission availability rate for future operations. In the course of sustained operations, aircraft maintenance must be carefully considered and programmed so that heavy flying requirements will not cause a continual decrease in aircraft availability.

Supported unit commanders, staffs, and logistical planners must conserve the use of available aircraft by—

- Establishing acceptable availability rates prior to operational commitment.
- Establishing FARPs to eliminate flying hours expended for those purposes.
- Utilizing surface means for transportation for logistical support whenever possible.
- Coordinating logistical planning to ensure full utilization of all aircraft sorties and to avoid duplication of effort.

Capabilities and Limitations

Capabilities

Marine Corps helicopters have the following capabilities:

- Under normal conditions, helicopters can ascend and descend at steep angles, a capability that enables them to operate from confined and unimproved areas.
- Troops and their combat equipment can be unloaded from a helicopter hovering a short distance above the ground with fast ropes and rappelling means, or if they can hover low

enough, the troops may jump to the ground. A rope ladder can be used to load personnel when the helicopter cannot land.

- Cargo can be transported as an external load and delivered to areas inaccessible to other types of aircraft or ground transportation.
- Because of a wide speed range and high maneuverability at slow speeds, helicopters can fly safely and efficiently at a low altitude, using terrain and trees for cover and concealment.
- A helicopter's ability to fly at high or low altitudes and to decelerate rapidly, combined with its capacity for slow forward speed and nearly vertical landing, enables it to operate under marginal weather conditions.
- Helicopters can land on the objective area in a tactical formation, LZ(s) permitting.
- Night/limited visibility landing and lift-offs can be made with a minimum of light.
- Helicopters flying at low levels are capable of achieving surprise, deceiving the enemy at the LZ(s), and employing shock effect through the use of suppressive fires.
- Engine and rotor noise may deceive the enemy as to the direction of approach and intended flight path.

Limitations

Limitations of Marine Corps helicopters include—

- The high fuel consumption rate of helicopters imposes limitations on range and aircraft load. Helicopters may reduce fuel load to permit an increased aircraft load. However, reducing the fuel load reduces the range and flexibility factors, which must be considered in planning.
- Weight and balance affect flight control. Loads must be properly distributed to keep the center of gravity within allowable limits.
- Hail, sleet, icing, heavy rains, and gusty winds (45 knots or more) limit or preclude use of helicopters.
- Engine/rotor noise may compromise secrecy.
- Aviator fatigue requires greater consideration in the operation of rotary-wing aircraft than in the operation of fixed-wing aircraft.
- The load carrying capability of helicopters decreases with increases of altitude, humidity, and temperature. This limitation may be compensated for though fuel load reduction.
- Crosswinds may affect the selection of direction of landing and lift off.

Characteristics

CH-46E	
Mission:	Assault troop transport
Alternate missions:	Cargo transport, casualty evacuation, TRAP
Crew configuration:	2 pilots, 1 crew chief/gunner, 1 aerial observer/gunner
Maximum speed:	145 knots indicated airspeed (KIAS)
Maximum endurance:	2 + 55 hours (hrs)
Weapons systems:	2.50 caliber XM 218 machine guns
Payload:	4,300 lbs (18 combat-loaded troops)

CH-53D	
Mission:	Assault transport of equipment and supplies
Alternate missions:	Assault troop transport, casualty evacuation, TRAP
Crew configuration:	2 pilots, 1 crew chief/gunner, 1 aerial observer/gunner
Maximum speed:	130 KIAS
Maximum endurance:	5 + 30 hrs
Air refuelable:	No (can offload fuel to FARP and aircraft on the ground)
Weapons systems:	2.50 caliber XM 218 machine guns
Payload:	37 passengers 24 combat-loaded troops @ 250 lbs = 6,000 lbs with 5 + 30 hr endurance Cargo - 13,000 lbs

CH-53E	
Mission:	Assault transport of heavy weapons, equipment, and supplies
Alternate missions:	Assault troop transport , casualty evacuation, TRAP
Crew configuration:	2 pilots, 1 crew chief/gunner, 1 aerial observer/gunner
Maximum speed:	150 KIAS
Maximum endurance:	4+00+ hrs
Air refuelable:	Yes (can offload fuel to FARP or aircraft on the ground)
Weapons systems:	2 .50 caliber XM 218 machine guns
Lift capability:	37 to 55 passengers 24 troops @ 250 lbs =6,000 lbs with 4+00+ hr endurance Cargo - 32,000 lbs

AH-1W	
Mission:	Fire support
Alternate missions:	FAC(A), TAC(A), escort, aerial reconnaissance
Crew:	2 pilots
Maximum speed:	170 KIAS
Maximum endurance:	2+30 hrs
Weapons systems:	Missiles: TOW, Hellfire, Sidearm, Sidewinder Guns: 20mm Rockets: 2.75 inch (7 or 19 shot pod) 5 inch (4 shot pod)

UH-1N	
Mission:	Utility support
Alternate missions:	Command and control, CAS, FAC(A), TAC(A), casualty evacuation, aerial reconnaissance, escort, assault troop transport
Crew configuration:	2 pilots, 1 crew chief/gunner, 1 aerial observer/gunner
Maximum speed:	130 KIAS (100 KIAS combat configured)
Maximum endurance:	+30 hrs (2+30 with auxiliary fuel bag)
Weapons systems:	2.75" rockets, 7.62mm miniguns, M240G, .50-caliber machine guns
Lift capability:	4 combat loaded troops @ 250 lbs = 1,000 lbs with 1+30 hr endurance

APPENDIX I

SAMPLE HELICOPTERBORNE WARNING ORDER

1. Situation

a. Enemy Forces

b. Friendly Forces

Co ___ likely to move by helo.

Co ___ likely to move by helo.

Co ___ likely to move by helo.

______ likely to move by ______.

______ likely to move by ______.

2. Probable Mission

My commander's intent is ___________________________________.

My intent is ____.

3. General Instructions

Anticipated PZs	________	________	________
	________	________	________
Anticipated LZs	________	________	________
	________	________	________
Anticipated objectives	_______	________	________
	________	________	________

Anticipated helicopter availability:

Co _____, _____ CH-46, _____, _____ CH-53D, _____ CH-53-E,

_____ UH-1N, Anticipated number of lifts: _____

Co _____, ______ CH- 46, ______, _____ CH-53D, _____ CH-53E,

_____UH-1N, Anticipated number of lifts: _____

4. Special Instructions

PZCO(s) __________ ______________________________

PZCO(s) __________ ______________________________

HST requests from landing support battalion:

Marines __________

Equipment __________ ______________________________

____

APPENDIX J

GLOSSARY

SECTION 1. ACRONYMS AND ABBREVIATIONS

ACE . . . . aviation combat element
AFL. assault flight leader
AMC . . . . air mission commander
AR. . . . . aerial refueling
ARVPZ . . . . aerial rendezvous pickup zone
ASC(A) . . . assault support coordinator (airborne)

BA. . . . . basic allowance

C2 .command and control
CAS. . . . . .close air support
CEOI . . . . communications-electronic operating instructions
Co . . . . .company
CSS . . . . combat service support
CSSD . . . . . .combat service support detachment
CSSE. . . . . .combat service support element

DASC . . . . direct air support center
DOA . . . . .day(s) of ammunition
DOS . . . . days of suppy

EFL. . . . . escort flight leader
EW . . . . .electronic warfare

FAC. . . . . .forward air controller
FAC(A). . . . . . . forward air controller (airborne)
FARP . . . . . . forward arming and refueling point
FEBA . . . . .forward edge of the battle area
FLOT . . . . .forward line of own troops
FM. . . . . frequency modulation
FMFM. . . . . Fleet Marine Force manual
FSC . . . . fire support coordinator
FSCC. . . . . .fire support coordination center
FSSG. . . . . force service support group

G-3 . . . . .operations staff officer (brigade or higher staff)
GCE . . . . ground combat element

HEALT . . . . .helicopter employment and assault landing table
HIMARS. . . . . high mobility artillery rocket system
HST. . . . . helicopter support team
HTF. . . . . helicopterborne task force
HUC . . . . helicopterborne unit commander
HWSAT . . . . helicopter wave and serial assignment table

IADS. . . . . Integrated Air Defense System
IPB . . . . intelligence preparation of the battlespace
ITG . . . . initial terminal guidance

JP. . . . . .joint publication

KIAS. . . . . .knots indicated airspeed
km . . . . kilometer

LAAD . . . . low altitude air defense
lb . . . . pound
LZ . . . . .landing zone

MAGTF . . . . Marine air-ground task force
MCRP . . . . . .Marine Corps reference publication
MCWP . . .Marine Corps warfighting publication
MEB . . . . Marine expeditionary brigade
MEF . . . . Marine expeditionary force
METT-T . . . . mission, enemy, terrain and weather, troops and support available—time available
MEU . . . . Marine expeditionary unit
MHE . . . . materials handling equipment
MLRS . . . . Multiple Launch Rocket System
mm . . . . .millimeter

NBC . . . . .nuclear, biological, and chemical
NSN . . . . National Stock Number

NVD . . . . night vision device
NWP . . . . .naval warfare publication

OCOKA-W . . . . observation and fields of fire, cover and concealment, obstacles, key terrain, avenues of approach, weather
OPORD. .operation order
OPSEC . . . . operations security

POW . . . . prisoner of war
PZ . . . . pickup zone
PZCO . . . . pickup zone control officer

S-1. . . . . manpower/personnel staff officer (battalion or regiment)
S-2. . . . . intelligence staff officer (battalion or regiment)
S-3. . . . . operations staff officer (battalion or regiment)
S-4. logistics staff officer (battalion or regiment)

SAM . . . . surface-to-air missile
SEAD . . . . . . .suppression of enemy air defenses
SOP. . . . . standing operating procedure

TAC(A). . . . . . tactical air coordinator (airborne)
TACP . . . . tactical air control party
TAR . . . . tactical air request
TDAR . . . . tactical defense alert radar
TOW . . . . tube launched, optically tracked wire guided missile
TRAP . . . . .tactical recovery of aircraft and personnel

UHF . . . . ultrahigh frequency

VHF . . . . very high frequency

WV . . . . wave

XO. . . . . .executive officer

SECTION 2. DEFINITIONS

acquisition—The process of locating a target with a search radar such that a tracking radar can take over and begin tracking the target. (To be incorporated into the next edition of MCRP 5-12C.)

air defense artillery—Weapons and equipment for actively combatting air targets from the ground. Also called ADA. (JP 1-02)

air mission commander—A mission commander, who shall be a properly qualified naval aviator or naval flight officer, should be designated when separate aircraft formations, each led by its own formation leader, are required for a common support mission or whenever a formation of four or more aircraft must perform a multiple sortie mission. The mission commander shall direct a coordinated plan of action and shall be responsible for the effectiveness of the mission. Also called AMC. (MCRP 5-12C)

air officer—An officer (aviator/naval flight officer) who functions as chief adviser to the commander on all aviation matters. An air officer is normally found at battalion level and higher within the ground combat element and within the Marine air-ground task force command element and combat service support element headquarters staffs. The air officer is the senior member of the tactical air control party. The battalion air officer supervises the training and operation of the two battalion forward air control parties. Also called AO. (MCRP 5-12C)

airspace coordination area—A three-dimensional block of airspace in a target area, established by the appropriate ground commander, in which friendly aircraft are reasonably safe from friendly surface fires. The airspace coordination area may be formal or informal. Also called ACA. (JP 1-02)

artillery preparation—Artillery fire delivered before an attack to disrupt communications and disorganize the enemy's defense. (AAP-6[2004])

assault support—The use of aircraft to provide tactical mobility and logistic support for the MAGTF, the movement of high priority cargo and personnel within the immediate area of operations, in-flight refueling, and the evacuation of personnel and cargo. (MCRP 5-12C)

assault support coordinator (airborne)—An aviator who coordinates, from an aircraft, the movement of aviation assets during assault support operations. Also called ASC(A). (MCRP 5-12C)

assault support helicopter—A helicopter which moves assault troops, equipment, and cargo into an objective area and which provides helicopter support to the assault forces.

battle position—(Army) 1. A defensive location oriented on a likely enemy avenue of approach. (FM3-90) 2. For attrack helicopters, an area designated in which they can maneuver and fire into a designated engagement area or engage targets of opportunity. (FM 1-112) (Marine Corps) 1. In ground operations, a defensive location oriented on an enemy avenue of approach from which a unit may defend. 2. In air operations, an airspace coordination area containing fire points for attack helicopters. Also called **BP**. (MCRP 5-12)

casualty evacuation—The movement of casualties. It includes movement both to and between-medical treatment facilities. Any vehicle may be used to evacuate casualties. Also called **CASE-VAC**. (JP 1-02). The movement of the sick, wounded, or injured. It begins at the point of injury or the onset of disease. It includes movement both to and between medical treatment facilities. All units have an evacuation capability. Any vehicle may be used to evacuate casualties. If a medical vehicle is not used it should be replaced with one at the first opportunity. Similarly, aeromedical evacuation should replace surface evacuation at the first opportunity. (MCRP 5-12C)

checkpoint—Geographical location on land or water above which the position of an aircraft in flight may be determined by observation or by electrical means. (Joint Pub 1-02) (Part three of four-part definition.)

close air support—Air action by fixed– and rotary-wing aircraft against hostile targets that are in close proximity to friendly forces and that require detailed integration of each air mission with the fire and movement of those forces. Also called **CAS**. (JP 1-02)

combat air patrol—An aircraft patrol provided over an objective area, the force protected, the critical area of a combat zone, or in an air defense area, for the purpose of intercepting and destroying hostile aircraft before they reach their targets. Also called **CAP**. (JP 1-02)

command and control aircraft—A tactical mission aircraft for the use of the helicopter coordinator (airborne) and helicopterborne unit commander to coordinate and control tactical helicopter assaults, troop movement, commander's reconnaissance, and other related missions.

communications-electronics operations instruction—An instruction containing details on call sign assignments, frequency assignments, codes and ciphers, and authentication tables and their use. The communications-electronic operating instructions (CEOI) is designated to complement information contained in operational unit communication standing operating procedures or Annex K (Combat Information Systems) to the operation order. The most common version if CEOI in use by the Marine Corps is the automated communications-electronics instructions, produced by the National Aeronautics and Space Administration. Also called **CEOI**.

control point—A position marked by a buoy, boat, aircraft, electronic device, conspicuous terrain feature, or other identifiable object which is given a name or number and used as an aid to navigation or control of ships, boats, or aircraft. (Joint Pub 1-02) (Part two of a four-part definition.)

day(s) of ammunition—Unit of measurement of replenishing ammunition expressed as a specified number of rounds, or items of bulk ammunition as may be appropriate per weapon, unit, individual kit, set, or using device required for one day of combat. Also called DOA. (MCRP 5-12C)

D-day—The unnamed day on which a particular operation commences or is to commence. (JP 1-02)

departure point—1. A navigational check point used by aircraft as a marker for setting course. 2. In amphibious operations, an air control point at the seaward end of the helicopter approach lane system from which helicopter waves are dispatched along the selected helicopter approach lane to the initial point. (JP 1-02)

direct air support center—The principal air control agency of the US Marine air command and control system responsible for the direction and control of air operations directly supporting the ground combat element. It processes and coordinates requests for immediate air support and coordinates air missions requiring integration with ground forces and other supporting arms. It normally collocates with the senior fire support coordination center within the ground combat element and is subordinate to the tactical air command center. Also called DASC. (JP 1-02)

effective range—That range at which a weapon or weapons system has a fifty percent probability of hitting a target (MCRP 5-12C)

electronic attack—See electronic warfare.

electronic intelligence—Technical and geolocation intelligence derived from foreign non-communications electromagnetic radiations emanating from other than nuclear detonations or radioactive sources. Also called ELINT. (JP 1-02)

electronic protection—See electronic warfare.

electronic warfare—Any military action involving the use of electromagnetic and directed energy to control the electromagnetic spectrum or to attack the enemy. Also called **EW**. The three major subdivisions within electronic warfare are: electronic attack, electronic protection, and electronic warfare support. a. electronic attack. That division of electronic warfare involving the use of electromagnetic energy, directed energy, or antiradiation weapons to attack personnel, facilities, or equipment with the intent of degrading, neutralizing, or destroying enemy combat capability and is considered a form of fires. Also called **EA**. EA includes: 1) actions taken to prevent or reduce an enemy's effective use of the electromagnetic spectrum, such as jamming and electromagnetic deception, and 2) employment of weapons that use either electromagnetic or directed energy as their primary destructive mechanism (lasers, radio frequency weapons, particle beams). b. electronic protection. That division of electronic warfare involving passive and active means taken to protect personnel, facilities, and equipment from any effects of friendly or enemy employment of electronic warfare that degrade, neutralize, or destroy friendly combat capability. Also called **EP**. c. electronic warfare support. That division of electronic warfare involving actions tasked by, or under direct control of, an operational commander to search for, intercept, identify, and locate or localize sources of intentional and unintentional radiated electromagnetic energy for the purpose of immediate threat recognition, targeting, planning and conduct of future operations. Thus, electronic warfare support provides information required for decisions involving electronic warfare operations and other tactical actions such as threat avoidance, targeting, and homing. Also called **ES**. Electronic warfare support data can be used to produce signals intelligence, provide targeting for electronic or destructive attack, and produce measurement and signature intelligence. (JP 1-02)

firing positions—In helicopterborne operations, a position occupied by an individual attack helicopter in order to engage targets. A battle position contains one or more firing positions. See also **battle position**.

fire support coordination center—A single location in which are centralized communications facilities and personnel incident to the coordination of all forms of fire support. Also called **FSCC**. (JP 1-02)

flight leader—A pilot qualified in model or helicopter aircraft commander designated in writing by the helicopter unit commander. In flight, including escorts, the overall flight leader is the helicopter transport commander.

forward air controller—An officer (aviator/pilot) member of the tactical air control party who, from a forward ground or airborne position, controls aircraft in close air support of ground troops. (JP 1-02) Also called **FAC**. (MCRP 5-12C)

forward air controller (airborne)—A specifically trained and qualified aviation officer who exercises control from the air of aircraft engaged in close air support of ground troops. The forward air controller (airborne) is normally an airborne extension of the tactical air control party. Also called **FAC(A)**. (JP 1-02)

forward arming and refueling point—A temporary facility–organized, equipped, and deployed by an aviation commander, and normally located in the main battle area closer to the area where operations are being conducted than the aviation unit's combat service area–to provide fuel and ammunition necessary for the employment of aviation maneuver units in combat. The forward arming and refueling point permits combat aircraft to rapidly refuel and rearm simultaneously. Also called **FARP**. (JP 1-02)

forward edge of the battle area—The foremost limits of a series of areas in which ground combat units are deployed, excluding the areas in which the covering or screening forces are operating, designated to coordinate fire support, the positioning of forces, or the maneuver of units. Also called **FEBA**. (JP 1-02)

forward line of own troops—A line that indicates the most forward positions of friendly forces in any kind of military operation at a specific time. The forward line of own troops (FLOT) normally identifies the forward location of covering and screening forces. The FLOT may be at, beyond, or short of the forward edge of the battle area. An enemy FLOT indicates the forward-most position of hostile forces. Also called **FLOT**. (JP 1-02)

ground fire—Small arms ground-to-air fire directed against aircraft. (JP 1-02)

helicopter assault force—A task organization combining helicopters, supporting units, and helicopterborne troop units for use in helicopterborne assault operations. (JP 1-02)

helicopterborne assault—The landing of helicopterborne forces within or adjacent to an objective area for the purpose of occupying and controlling the objective area and positioning units for action against hostile forces.

helicopterborne operation—A military action in which combat forces and their equipment maneuver about the battlefield by helicopters or vertical-landed aircraft. (MCRP 5-12C)

helicopterborne unit commander—The ground officer who has been designated by the MAGTF commander to be the commander of the helicopterborne force and who is charged with the accomplishment of the ground tactical plan. Also called **HUC**.

helicopter direction center—In amphibious operations, the primary direct control agency for the helicoter group/unit commander operating under the overall control of the tactical air control center. (JP 1-02) The helicopter direction center is an agency within the Navy tactical air control system and is position afloat. The helicopter direction center is not a Marine air command and control system agency, but it interacts closely with the direct air support center in the control of helicopter operations between ship and shore. The helicopter direction center also interacts closely with the air support element of the Marine expeditionary unit aviation combat element. Also called **HDC**. (MCRP 5-12C)

helicopter employment and assault landing table—A planning document prepared jointly by the helicopter and helicopterborne unit commanders. It includes detailed plans for the movement of helicopterborne troops, equipment, and supplies. It is the landing timetable for the helicopter movement uniting scheduled units with numbered flights and waves and provides the basis for the helicopter unit's flight schedule. It is used by the appropriate air control agency to control the helicopter movement. Also called **HEALT**.

helicopter support team—1. A task organization formed and equipped for employment in a landing zone to facilitate the landing and movement of helicopterborne troops, equipment, and supplies, and to evacuate selected casualties and enemy prisoners of war. Also called **HST**. (JP 1-02) 2. Within the Marine Corps, helicopter support teams are sourced fromthe force service support group, specifically from the landing support company of the support battalion. (To be included in the next edition of MCRP 5-12C.)

helicopter wave—One or more helicopters grouped under a single leader scheduled to land in the same landing zone at approximately the same time. A helicopter wave is composed of one or more flights and can consist of helicopters from more than one ship. See also wave.

helicopter wave and serial assignment table—A planning document utilized in helicopterborne operations describing the tactical unit, equipment, and supplies that are to be loaded into each helicopter. The table identifies each heliteam with its assigned serial number and the serial number with the flight and wave. Also called **HWSAT**.

H-hour—The specific hour on D-day at which a particular operation commences. (JP 1-02)

high frequency—A high frequency of 3 MHz to 30 MHz.

holding point—A geographically or electronically defined location used in stationing aircraft in flight in a predetermined pattern in accordance with air traffic control clearance. (JP 1-02)

hostile area—Area of known enemy concentration in which intense opposition can be expected. It differs from an insecure area in that no friendly forces are in the immediate area, landing zones are unprotected, and fixed-wing preparation fire is normally mandatory.

initial point—1. The first point at which a moving target is located on a plotting board. 2. A well-defined point, easily distinguishable visual and/or electronically, used as a starting point for the bomb run to the target. 3. **airborne**-A point close to the landing area where serials (troop carrier air formations) make final alterations in course to pass over individual drop zone or landing zones. 4. **helicopter**-An air control point in the vicinity of the landing zone from which individual flights of helicopters are directed to their prescribed landing sites. 5. Any designated place at which a column or element thereof is formed by the successive arrival of its various subdivisions, and comes under the control of the commander ordering the move. (JP 1-02)

initial terminal guidance—A mission normally assigned to reconnaissance units to provide the helicopter coordinator (airborne) with information resulting from prelanding reconnaissance. They establish and operate signal devices for guiding the initial helicopter waves from the initial point to the landing zone. Also called **ITG**.

intelligence requirement—1. Any subject, general or specific, upon which there is a need for the collection of information or the production of intelligence. 2. A requirement for intelligence to fill a gap in the command's knowledge or understanding of the battlespace or threat forces. (JP 1-02) 3. In Marine Corps usage, questions about the enemy and the environment, the answers to which a commander requires to make sound doctrine. Also called **IR**. (MCRP 5-12C)

Jamming—The deliberate radiation or reflection of electromagnetic energy to prevent or degrade the receipt of information by a receiver. It includes communications and noncommunications jamming. (MCRP 5-12A)

laager point—Secure location on the ground designated by aviation units utilized for the rendezvous, marshalling, or positioning of flights of aircraft between missions or awaiting completion or activation of an assigned mission. Other than communications, no other support should be required. This site may be isolated and independent or it may be adjacent to an airfield, facility, or forward arming and refueling point.

landing point—A point within a landing site where one helicopter or vertical takeoff and landing aircraft can land. (JP 1-02)

landing site—A site within a landing zone containing one or more landing points. (JP 1-02)

landing zone—A specified zone used for the landing of aircraft. (JP 1-02)

L-hour—In amphibious operations, the time at which the first helicopter of the helicopterborne assault wave touches down in the landing zone. (MCRP 5-12C)

low frequency—A frequency of 30 kHz to 300 kHz.

low level flight—Flight conducted at constant airspeed and indicated altitude at which detention or observation of an aircraft or of the points from which and to which it is flying is avoided or minimized. The route is preselected and conforms generally to a straight line.

Marine air-ground task force—The Marine Corps principal organization for all missions across the range of military operations, composed of forces task-organized under a single commander capable of responding rapidly to a contingency anywhere in the world. The types of forces in the Marine air-ground task force (MAGTF) are functionally grouped into four core elements: a command element, an aviation combat element, a ground combat element, and a combat service support element. The four core elements are categories of forces, not formal commands. The basic structure of the MAGTF never varies, though the number, size, and type of Marine Corps units comprising each of its four elements will always be mission dependent. The flexibility of the organizational structure allows for one or more subordinate MAGTFs, other Service and/or foreign military forces, to be assigned or attached. Also called **MAGTF**. (MCRP 5-12C)

mission brief—The final phase of the planning effort that should include, as attendees, all mission participants. This brief will set forth the concept of operations, ground tactical plan, scheme of maneuver from the pickup zone through the objective, and specific details concerning mission, coordination, and execution.

mission, enemy, terrain and weather, troops and support available–time available—Factors to be considered in estimating the situation during the planning of the military operation.

pickup zone—The zone in which helicopters land to pick up troops and supplies for movement to the landing zone.

pickup zone control officer—An officer who organizes, controls, and coordinates operations in pickup zones.

priority intelligence requirements—1. Those intelligence requirements for which a commander has an anticipated and stated priority in the task of planning and decisionmaking. (JP 1-02) 2. In Marine Corps usage, an intelligence requirement associated with a decision that will critically affect the overall success of the command's mission. Also called PIR. (MCRP 5-12C)

secure area—An area that has not received hostile fire for 72 hours and in which helicopters will most likely not be subject to fire during the approach, landing, takeoff, and departure.

suppressive fire—Fires on or about a weapons system to degrade its performance below the level needed to fulfill its mission objectives, during the conduct of the fire mission. (JP 1-02)

tactical air command center—The principal Marine Corps air command and control agency from which air operations and air defense warning functions are directed. It is the senior agency of the Marine air command and control system which serves as the operational command post of the aviation combat element commander. It provides the facility from which the aviation combat element commander and his battlestaff plan, supervise, coordinate, and execute all current and future air operations in support of the Marine air-ground task force. The tactical air command center can provide integration, coordination, and direction of joint and combined air operations. Also called **Marine TACC**. (MCRP 5-12C)

tactical air control party—A subordinate operational component of a tactical air control system designated to provide air liaison to land forces and for the control of aircraft. (JP 1-02) In the Marine Corps, tactical air control parties are organic to infantry divisions, regiments, and battalions. Tactical air control parties establish and maintain facilities for liaision and communications between parent units and airspace control agencies, inform and advise the ground unit commander on the employment of supporting aircraft, and request and control air support. Also called **TACP**. (MCRP 5-12C)

tactical air coordinator (airborne)—An officer who coordinates, from an aircraft, the action of combat aircraft engaged in close support of ground or sea forces. (JP 1-02) Within the Marine air command and control system, the tactical air coordinator (airborne) is the senior air coordinator having authority over all aircraft operating within his assigned area. The tactical air coordinator (airborne), considered an airborne extension of the direct air support center and fire support coordination center, contributes to coordination among the tactical air control parties, airborne forward air controllers, and the fire direction of artillery and naval gunfire. Also called **TAC(A)**. (MCRP 5-12C)

ultrahigh frequency—A frequency of 300 to 3,000 MHz.

very high frequency—A frequency of 30 to 300 MHz.

wave—A formation of forces, landing ships, craft, amphibious vehicles or aircraft, required to beach or land about the same time. Can be classified as to type, function or order as shown: a. assault wave; b. boat wave; c. helicopter wave; d. numbered wave; e. on-call wave; f. scheduled wave. (JP 1-02)

APPENDIX K

REFERENCES

Joint Publications (JP)

1-02	Department of Defense Dictionary of Military and Associated Terms
3-02.1	Joint Doctrine for Landing Force Operations

Navy Warfare Publication (NWP)

3-22.5-CH46E	CH-46E Tactical Manual
3-22.5-HH60H	HH-60H Tactical Manual
3-22.5-MH53 PG	MH-53 Tactical Manual Pocket Guide
3-22.5-S3B	S-3B Tactical Manual
3-22.5-SAR-TAC	Navy Search and Rescue Tactical Information Document (SAR TACAID)

Marine Corps Warfighting Publications (MCWPs)

3-11.2	Marine Rifle Squad (under development)
3-16	Fire Support Coordination in the Ground Combat Element
3-25.10	Low Altitude Air Defense Handbook
3-40.3	Communications and Information Systems

Marine Corps Reference Publications (MCRPs)

2-12A	Intelligence Preparation of the Battlespace
4-11.3E	Multiservice Helicopter Sling Load, Volumes I, II, and III
5-2A	Operational Terms and Graphics
5-12C	Marine Corps Supplement to the Department of Defense Dictionary of Military and Associated Terms

Made in the USA
Monee, IL
04 November 2026

67382861R10075